Abdo Gaber

Wireless Indoor Positioning

Abdo Gaber

Wireless Indoor Positioning

based on TDOA and DOA Estimation Techniques using IEEE 802.11 Standards

Südwestdeutscher Verlag für Hochschulschriften

Impressum / Imprint
Bibliografische Information der Deutschen Nationalbibliothek: Die Deutsche Nationalbibliothek verzeichnet diese Publikation in der Deutschen Nationalbibliografie; detaillierte bibliografische Daten sind im Internet über http://dnb.d-nb.de abrufbar.
Alle in diesem Buch genannten Marken und Produktnamen unterliegen warenzeichen-, marken- oder patentrechtlichem Schutz bzw. sind Warenzeichen oder eingetragene Warenzeichen der jeweiligen Inhaber. Die Wiedergabe von Marken, Produktnamen, Gebrauchsnamen, Handelsnamen, Warenbezeichnungen u.s.w. in diesem Werk berechtigt auch ohne besondere Kennzeichnung nicht zu der Annahme, dass solche Namen im Sinne der Warenzeichen- und Markenschutzgesetzgebung als frei zu betrachten wären und daher von jedermann benutzt werden dürften.

Bibliographic information published by the Deutsche Nationalbibliothek: The Deutsche Nationalbibliothek lists this publication in the Deutsche Nationalbibliografie; detailed bibliographic data are available in the Internet at http://dnb.d-nb.de.
Any brand names and product names mentioned in this book are subject to trademark, brand or patent protection and are trademarks or registered trademarks of their respective holders. The use of brand names, product names, common names, trade names, product descriptions etc. even without a particular marking in this work is in no way to be construed to mean that such names may be regarded as unrestricted in respect of trademark and brand protection legislation and could thus be used by anyone.

Coverbild / Cover image: www.ingimage.com

Verlag / Publisher:
Südwestdeutscher Verlag für Hochschulschriften
ist ein Imprint der / is a trademark of
OmniScriptum GmbH & Co. KG
Heinrich-Böcking-Str. 6-8, 66121 Saarbrücken, Deutschland / Germany
Email: info@svh-verlag.de

Herstellung: siehe letzte Seite /
Printed at: see last page
ISBN: 978-3-8381-5143-4

Zugl. / Approved by: Magdeburg, Otto-von-Guericke-Universität Magdeburg, Diss., 2015

Copyright © 2015 OmniScriptum GmbH & Co. KG
Alle Rechte vorbehalten. / All rights reserved. Saarbrücken 2015

Wireless Indoor Positioning based on TDOA and DOA Estimation Techniques using IEEE 802.11 Standards

Abdo Gaber

Dedications

This work is dedicated to the memory of my mother, my father, my children Ahmed and Safwan, and my wife

Abdo Gaber

Acknowledgements

I would like to express my sincere thanks and gratitude to all those gave me the support to complete this work.

First of all I want to thank my advisor Prof. Dr.-Ing. Abbas Omar for a lot more than his help, support, and guidance, and for a lot more than skills and knowledge that I learned from him through all the years of my PhD. His insight, experience, and knowledge provided me the required motivation to complete this work.

I express my appreciation to the German Academic Exchange Service (DAAD) for the financial support during my study in Germany. At the same time, I would like to thank Otto von Guericke University of Magdeburg to provide the opportunity and support me to carry out this research.

I would like to express my thanks to Prof. Dr.-Ing. Christian Diedrich and Prof. Dr. Nuno Borges Carvalho (University of Aveiro, Portugal) to be the reviewers of my dissertation, furthermore, I am also thankful to Prof. Dr.-Ing. Andreas Lindemann and Prof. Dr.-Ing. Edmund P. Burte to be part of the examination committee for this dissertation.

I would like to extend my sincere thanks to current and former members of Institute for Information Technology and Communications namely Prof. Dr. Al-Hamadi, Dr. Jstingmeier, Mr. Winkler, Mr. Ulrich, Dr. Bresch, Mr. Dempwolf, and Dr. Panzner, who have provided hearty help in many different ways and shared numerous pleasant hours. I am expressing my sincere gratitude to my close friends Moftah, Anwar, Ali, Mustafa, Ibrahim, and many friends in Magdeburg for their help, support, and more importantly the great and memorable time that we spent with each other.

Words cannot express my gratefulness to my parents, my brothers and my sisters for their care and support, whose love is the source of my happiness and strength. Finally, I would like to thank my wife Karama. Her love, effort, patience, and understanding are the source of my motivation to continue.

Abdo Gaber
Magdeburg, Germany
May 7, 2015.

Zusammenfassung

Drahtlose Ortung mit Hilfe von Funksignalen hat eine erhebliche Bedeutung in den Bereichen von Navigation und Zielverfolgung eingenommen. Jedoch können die zurzeit existierenden und von Satelliten abhängigen Navigationssysteme aufgrund Abschattungen und der vielfältigen Quellen von Störungen und Interferenz in urbanen und geschlossenen Räumen (sogenannte Indoor - Umgebungen) keine genaue Positionsschätzung bieten. Die drahtlose Ortung unterscheidet sich stark von traditionellen Systemen, besonders im Hinblick auf Betriebsumgebung, Systembedingungen und die erforderliche Anwendung einschließlich der erforderlichen Genauigkeit. Daher werden neue Ortungssysteme gebraucht, die spezifisch für Indoor- Ortungsanwendungen bestimmt sind und die Fähigkeit haben, die Herausforderungen der Indoor-Umgebung zu überwin-den. Infolgedessen stellt sich die Indoor-Ortung als ein neues wichtiges Forschungsgebiet heraus. Sie wird in Zukunft in zahlreichen Anwendungen benutzt werden.

In dieser Arbeit wird eine effektive drahtlose Methode der Indoor-Ortung entwickelt, welche auf vorhandenen Signalen des IEEE 802.11 Standards wie 802.11a, 802.11n und 802.11ac aufsetzt. In aktuellen drahtlosen Netzwerken werden Technologien wie MIMO-Antennen (Multiple Input Multiple Output) und OFDM (Orthogonal Frequency Division Multiplexing) genutzt, daher wird mit Hilfe dieser Technologien ein Framework für ein drahtloses Ortungssystem entwickelt und erforscht. Die Algorithmen sollen effizient und in Echtzeit auf einem digitalen Signalprozessor implementiert werden. Die Grundidee der Arbeit ist die Nutzung der Präambel von OFDM-Frames ohne Modifizierung, um den Kanalzustand für zusätzliche Zwecke neben der Demodulation der Datenanteile zu messen. Um die höchste Genauigkeit der Kanalparameter zu erzielen, wurden bestimmte Felder der Präambel ausgewählt.

Eine präzise Schätzung der Ankunfts-Zeit-Differenz (TDOA, Time Difference of Arrival) der Signale in einem Mehrweg-Kommunikationskanal sorgt für eine hohe Auflösung der drahtlosen Ortung. Das Problem der genauen Schätzung der Zeitverzögerung und der relativen Amplitude der Mehrweg-Signale wurde für die drahtlose Ortung anhand TDOA bewertet. Moderne Methoden, die auf Untervektorraum-Algorithmen (Matrix Pencil (MP), Unitary Matrix Pencil (UMP) und Beam-space Matrix Pencil (BMP)) basieren, wurden entwickelt und in verschiedenen Implementierungen realisiert, um die Parameter aus dem gemessenen Kanalfrequenzgang (CFR) mithilfe von OFDM-Systemen zu schätzen.

Da die aktuellen drahtlosen Netzwerke MIMO-OFDM-Technologie nutzen, kann die Verfü-gbarkeit von Antennen-Arrays in Basisstationen (BS) für die Schätzung der Direction of Arrival (DOA) als eine weitere Beobachtung und als eine Art räumliche Diversität für die TDOA-Schätzung verwendet werden. Desweiteren wird die OFDM-Frequenz-Diversität genutzt. Daher wird das Problem der genauen Schätzung der Zeitverzögerung, der relativen DOA und der relativen Amplitude von Mehrweg-Signalen für die drahtlose Ortung bewertet. Das Prinzip, wie man Multi-Antennen Multi-Carrier Systeme für drahtlose Ortung benutzt, wird behandelt. Die aktuellen Untervektorraum-Algorithmen (2-D MP, 2-D UMP und 2-D BMP) wurden in einer neuen Methodik angewandt, um die Parameter aus dem gemessenen Raum-Kanal-Frequenzgang mittels Antennen-Array und Breitband-OFDM-Signalen zu schätzen. Außerdem wird das Ausnutzen von Phänomenen wie Zeit-Diversität, Frequenz-Diversität und Raum-Diversität untersucht, um die Ortungssystemleistung zu verbessern.

Eine geschlossene analytische Lösung vom Schätzungsproblem für die MU-Koordinaten soll entwickelt werden. Ein nicht-iterativer gewichteter Kleinst-Quadrate-Schätzer (W-LS) wird für die TDOA-Schätzungen basierte MU-Koordinaten präsentiert. Weiterhin wird ein nicht iterativer hybrider W-LS-Schätzer basierend auf TDOA und DOA präsentiert. Die oben genannten Schätzer lösen das Problem vom Initialwert und das Problem vom Aufteilen der vorhandenen Schätzer. Sie sind robust gegenüber Kanalfading und nicht optimalem Signal-Rausch-Verhältnis (SNR). Um den Effekt der Fundamentalparameter des drahtlosen Ortungssystems auf die Ortungsleistung zu messen, wird die Cramer-Rao-Ungleichung (CRB, Cramer-Rao Bound) aus der Fehlervarianz der TDOA und der DOA-Schätzung abgeleitet.

Eine bekannte Herausforderung in der zeitbasierten drahtlosen Indoor-Ortung ist das Blockierungsproblem des direkten Weges zwischen Sender und Empfänger aufgrund der nicht optimalen Beschaffenheit von Indoor-Umgebungen. Das Fehlen eines direkten Weges führt zu einem sehr großen Abstands-Messfehler. Deshalb ist die Identifizierung des Status eines unbekannten direkten Weges (UDP, Undetected Direct Path) eine der wichtigsten Herausforderungen für die drahtlose Indoor-Ortung. Die Einbeziehung der Kanalstörung verbessert die Genaui-gkeit des Ortungssystems. Daher werden die genaue Schätzung der Kanalprofil-Parameter und die richtige Modellierung des erkannten direkten Weges (DDP, Detected Direct Path) und UDP Kanalprofile auf das Problem der Identifikation UDP gerichtet und behandelt. Die Werte des UPD-Status können dann ignoriert werden.

Um die Leistung der verschiedenen vorgeschlagenen Algorithmen zu bewerten, wurde eine Reihe von typischen Experimenten in Indoor-Umgebungen vorgenommen.

Schlüsselwörter: 1-D Matrix Pencil Algorithmen; 2-D Matrix Pencil Algorithmen; Cramer-Rao bound; DOA Schätzung; IEEE 802.11 Standards; MIMO-OFDM; Zeitverzögerung Schätz-ung; UDP Identifikation.

Abstract

Wireless positioning using radio signals has received considerable attention in the field of navigation and tracking. However, the existing navigation systems that are based on satellites cannot provide accurate position estimation in urban and indoor environments due to shadowing and various sources of noise and interference. In fact, wireless indoor positioning is very different from those traditional systems in many aspects such as the operating environments, system requirements, and the required applications including the required accuracy. Therefore, there is a need for new positioning systems that are specifically designed for indoor applications and can overcome the challenges of indoor environments. As a result, the wireless indoor positioning is emerging as a new important research area, its services will be widely used in the future through a number of applications.

In this work, an effective wireless indoor positioning has been developed based on the opportune signals of IEEE 802.11 standards such as 802.11a, 802.11n, and the emerging 802.11ac. The recent wireless networks are equipped with multiple input multiple output (MIMO) and orthogonal frequency-division multiplexing (OFDM) technologies. A framework for wireless indoor positioning using recent wireless networks needs to be developed and investigated. The developed algorithms should use a very efficient computational methodology for real time implementation on a digital signal processor (DSP) chip. The key element of our work is to use the preamble of the OFDM frames without any modification to measure the channel state for additional purposes to the demodulation of the data portion. The appropriate training fields have been selected to achieve the highest range of estimated channel profile parameters.

For high-resolution wireless positioning, the time difference of arrival (TDOA) associated with signals in a multipath communication channel should be estimated. The problem of highly resolving the propagation time delays and the relative amplitudes of multipath signals has been addressed for wireless positioning based on the TDOA observations. The recent subspace-based algorithms, represented by one-dimensional Matrix Pencil (1-D MP), 1-D Unitary Matrix Pencil (1-D UMP), and 1-D Beam-space Matrix Pencil (1-D BMP) algorithms, have been enhanced and implemented in different realizations to estimate those parameters from the measured channel frequency response (CFR) using OFDM signals.

Since the most-recent wireless networks use the MIMO-OFDM technology, the availability of antenna arrays in base stations (BSs) can be used to estimate the direction of arrival (DOA) as another observation and as a kind of spatial diversity for TDOA estimation besides the frequency diversity coming from OFDM. Therefore, the problem of highly resolving the propagation time delays, the relative DOAs, and the relative amplitudes of multipath signals has been addressed for a high-resolution wireless positioning system. The principle of how to use multi-antenna multi-carrier systems for wireless positioning is revealed. The recent subspace-based algorithms, represented by 2-D MP, 2-D UMP, and 2-D BMP algorithms, are applied in a new way to estimate these parameters simultaneously from the measured space channel frequency response (S-CFR) using multiple antennas and wideband orthogonal multi-carrier signals. In addition, the principle of using diversity techniques such as time diversity, frequency diversity, and space diversity has been investigated to improve the performance of the positioning system.

A closed-form solution to the estimation problem of mobile unit (MU) coordinates should be developed. A non-iterative weighted least square (W-LS) estimator is presented to estimate the MU coordinates based on the TDOA estimates. Furthermore, a non-iterative hybrid W-LS estimator has been presented based on the TDOA and DOA estimates. The proposed estimators solve the problem of the initial guess and that of partitioning of the other estimators. They are also robust against channel fading and low signal to noise ratio (SNR). To measure the effect of the fundamental parameters of a wireless positioning system on the positioning performance, the Cramer-Rao bound (CRB) of TDOA and DOA estimation error variances has been derived.

A well-known challenge in time-based wireless indoor positioning is the problem of obstruction of the direct path between the transmitter and the receiver due to the harsh nature of indoor environments. The absence of direct path leads to a very large distance measurement error (DME). Therefore, one of the major challenges for wireless indoor positioning is the identification of undetected direct path (UDP) channel profiles. Adding the channel obstruction knowledge improves the accuracy of the positioning system. Therefore, the accurate estimation of channel profile parameters and the proper modeling of detected direct path (DDP) and UDP channel profiles have been treated and addressed to the problem of UDP identification. The results of UDP condition can then be mitigated.

To evaluate the performance of various proposed algorithms, a number of experiments have been made in typical indoor environments.

Index Terms—1-D matrix pencil algorithms; 2-D matrix pencil algorithms; Cramer-Rao bound; DOA estimation; IEEE 802.11 standards; MIMO-OFDM; time delay estimation; UDP Identification.

Contents

Dedications	iii
Acknowledgements	v
Zusammenfassung	vii
Abstract	ix
Contents	xi
List of Figures	xiv
List of Tables	xix
Mathematical Notation	xxiii
List of Symbols	xxv
List of Abbreviations	xxix

1	**Introduction to Wireless Indoor Positioning**	**1**
1.1	Basic Observations	1
1.2	Wireless Indoor Positioning Applications	3
1.3	The Main Challenges of Wireless Indoor Radio Propagation Channel	4
1.4	Brief Literature Overview	5
1.5	Objectives of the Dissertation	7
1.6	The Dissertation Outline and Contributions	9
2	**Fundamentals of Position Estimation Techniques**	**13**
2.1	Wireless Positioning System Architecture	13
2.2	Indoor Radio Propagation Channel Model	14
	2.2.1 The CFR and CIR of Multipath Wireless Channel Understanding	15
	2.2.2 Time Delay Estimation using Channel Delay Profile	17
2.3	Transmitter Position Categories	19
2.4	Bandwidth Consideration	21
2.5	Main Sources of Error in Time-based Wireless Indoor Positioning Systems	23
2.6	Radio Signal Characteristics	24
	2.6.1 Received Signal Strength (RSS) based Techniques	24
	2.6.1.1 Path loss Model based Techniques	24
	2.6.1.2 Fingerprinting based Techniques	25
	2.6.2 Direction of Arrival (DOA) based Techniques	27
	2.6.3 Signal Propagation Time Delay based Techniques	29
	2.6.3.1 Indoor Positioning Techniques based on TOA	30
	2.6.3.2 2-D Wireless Positioning based on TOA	30

Contents

        2.6.3.3   Indoor Positioning Techniques based on TDOA 33

3 System Model and Time Delay Estimation using 1-D Matrix Pencil Algorithms   **37**
- 3.1 OFDM Signal Model 38
- 3.2 The symmetry of OFDM Time Delay and DOA Estimation Problems . . . 39
- 3.3 OFDM System Sensors 40
    - 3.3.1 IEEE 802.11a Frame Format 40
    - 3.3.2 IEEE 802.11n and IEEE 802.11ac Frame Formats 42
- 3.4 Super-Resolution Algorithms for Wireless Indoor Positioning 44
- 3.5 1-D Matrix Pencil Algorithm 45
- 3.6 1-D Unitary Matrix Pencil Algorithm 50
- 3.7 1-D Beam-space Matrix Pencil Algorithm 52
- 3.8 Diversity Techniques using 1-D Matrix Pencil Algorithms 55
- 3.9 Computational Complexity 56

4 Joint Time Delay and DOA Estimation using 2-D Matrix Pencil Algorithms   **59**
- 4.1 2-D Uniform Rectangular Array Problem Formulation 60
- 4.2 MIMO-OFDM System Model 60
    - 4.2.1 Uniform Linear Antenna Array Design 63
- 4.3 2-D Matrix Pencil Algorithm 64
    - 4.3.1 Correlation Maximization Pairing Method 70
    - 4.3.2 Proposed Pairing Method 70
- 4.4 2-D Unitary Matrix Pencil Algorithm 72
- 4.5 2-D Beam-space Matrix Pencil Algorithm 75
- 4.6 Diversity Techniques using 2-D Matrix Pencil Algorithms 78
- 4.7 Computational Complexity 79
- 4.8 Computational Complexity Comparison of 1-D and 2-D Matrix Pencil Algorithms 82

5 Mobile Unit Position Estimation Based on TDOA and DOA Measurements   **85**
- 5.1 Estimators Based on TDOA Measurements 85
    - 5.1.1 Iterative Least Square Estimator 86
    - 5.1.2 Divide-and-Conquer based on TDOA Estimator 87
    - 5.1.3 Chan Estimator 88
- 5.2 Estimators Based on DOA Measurements 90
    - 5.2.1 Lines Intersection Estimator 91
    - 5.2.2 Least Square Estimator 91
    - 5.2.3 Divide-and-Conquer based on DOA Estimator 92
    - 5.2.4 Stansfield-Least Square Estimator 93
- 5.3 Hybrid TDOA and DOA for Position Estimation 94
    - 5.3.1 Hybrid DAC based on TDOA and DOA Estimator 94
    - 5.3.2 Weighted TDOA and DOA Estimator 95
- 5.4 Proposed Hybrid TDOA and DOA Estimator 95
    - 5.4.1 Initial Position Estimation using Weighted-Least Square Estimator 96
    - 5.4.2 Final Position Estimation using Hybrid TDOA and DOA Estimator 98
- 5.5 Lower Bounds of TDOA and DOA Estimation Error Variances 99

Contents

	5.5.1	DOA Estimation Error Variance	101
	5.5.2	TDOA and Time Delay Estimation Error Variance	104
	5.5.3	Joint Time Delay and DOA Estimation Error Variance	108

6 Performance Evaluation Based on Channel Measurements — 117

- 6.1 Measurement Tools 118
  - 6.1.1 The Vector Signal Generator (Agilent MXG N5182A) 118
  - 6.1.2 The Antennas 118
  - 6.1.3 The Signal Analyzer (Agilent EXA N9010A) 118
  - 6.1.4 The Software Defined Radio Modules 119
  - 6.1.5 The Network Analyzer (Agilent ENA E5071C) 120
  - 6.1.6 The Wideband Time Domain Transmission (Agilent DCA 86100A) 120
- 6.2 RSS Techniques Investigation 121
  - 6.2.1 Investigation of Distance Estimation using Path Loss Model . . . 121
  - 6.2.2 Investigation of Wireless Positioning using Fingerprinting 122
- 6.3 Simulation Results 124
  - 6.3.1 Effect of Pencil Value 124
  - 6.3.2 Effect of SNR and Temporal and Frequency Diversities 126
- 6.4 Measurement Systems Using 1-D MP Algorithms 126
  - 6.4.1 Measurements Analysis using 20 MHz Bandwidth 126
  - 6.4.2 TDOA Estimation using a LOS Wireless Channel and a Reference Signal 132
  - 6.4.3 Time Delay Estimation using a LOS Wireless Channel with Diversity Techniques 132
  - 6.4.4 Time Delay Estimation using a NLOS Wireless Channel with Diversity Techniques 138
- 6.5 Measurement Systems Using 2-D MP Algorithms 141
  - 6.5.1 Joint Time Delay and DOA Estimation using a LOS Wireless Channel 141
  - 6.5.2 Joint Time Delay and DOA Estimation using a NLOS Wireless Channel 142
- 6.6 Measurement Analysis for 2-D Wireless Indoor Positioning 145
  - 6.6.1 Using 1-D MP Algorithms for 2-D Wireless Indoor Positioning . 147
    - 6.6.1.1 Position RMSE versus System Bandwidths 147
    - 6.6.1.2 Position RMSE versus Number of Antennas 148
  - 6.6.2 Using 2-D MP Algorithms for 2-D Wireless Indoor Positioning . 151
    - 6.6.2.1 Position RMSE versus System Bandwidths and Number of Antennas using TDOA Estimates 151
    - 6.6.2.2 Position RMSE versus System Bandwidths and Number of Antennas using DOA Estimates 152
    - 6.6.2.3 Position RMSE using Hybrid TDOA and DOA 154
      - 6.6.2.3.1 Position RMSE using DAC based on TDOA and DOA Estimator 158
      - 6.6.2.3.2 Position RMSE using Weighted TDOA and DOA Estimator 158

	6.6.2.3.3 Position RMSE using Proposed Hybrid TDOA and DOA Estimator 158

    6.6.3 Performance Comparison between 1-D MP with Spatial Diversity and 2-D MP Algorithms 161
    6.6.4 2-D Wireless Indoor Positioning at a Number of Static Positions . 163

7 UDP Identification for High-Resolution Wireless Indoor Positioning   169
7.1 Introduction to the Problem of UDP Channel 169
7.2 System Model 170
7.3 Statistical Modeling of Multipath Channel Features 171
    7.3.1 Mean Excess Time Delay 174
    7.3.2 Total Power 174
    7.3.3 Hybrid Time-Power Parameter 176
7.4 Likelihood-ratio Test for UDP Channel Profile Identification 176
7.5 System Performance 179

8 Conclusions and Future Work   185
8.1 Conclusions 185
8.2 Future Work 188

Appendices   191

A The Unitary Matrix Transformation   193

B Kronecker and Khatri-Rao Products   195

C The QR and QZ algorithms   197

Bibliography   199

List of Figures

1.1	The wireless indoor positioning applications.	4
1.2	Multipath radio environment.	6
1.3	Characteristics of wireless indoor radio propagation channel.	6
2.1	The general block diagram of a wireless positioning system based on the radio signal characteristics.	14
2.2	Time delay and DOA representation using a uniform linear antenna array.	16
2.3	CIR representation as a tapped delay module.	16
2.4	Measurement environment.	17
2.5	The CIR obtained using IFFT with 320 MHz BW.	18
2.6	The CFR of recorded channel comprising frequency selectivity.	18
2.7	The CIR obtained using IFFT with 160 MHz BW.	20
2.8	The CIR obtained using ICZT with 160 MHz BW.	20
2.9	The MU position categories based on the availability of the direct path.	22
2.10	Wireless positioning system based on distance estimation using RSS technique.	26
2.11	The summary of RSS measurement techniques.	27
2.12	Triangulation illustration using DOA technique in 2-D scenario.	28
2.13	2-D TOA positioning system using three receivers.	31
2.14	2-D TOA positioning system using four receivers.	33
2.15	The geometric relationship between a MU and three fixed BSs in a TDOA system.	35
3.1	The symmetry of OFDM time delay and DOA estimation problems [1].	41
3.2	OFDM training structure.	41
3.3	The spectrum of STS and LTS training sequences of 802.11a.	43
3.4	The transmitted preamble in the time-domain of 802.11a.	43
3.5	PPDU formats of 802.11n and 802.11ac.	45
3.6	Operational flow of the proposed TDOA estimation using 1-D MP algorithms.	56
3.7	Comparison of computational complexity of the initial transformation and SVD of various 1-D MP algorithms.	57
4.1	Signal modeling at the input of the URA.	61
4.2	The ULA and the virtual OFDM sensors representation	61
4.3	The designed ULA composing 8 elements.	65
4.4	The affect of ULA coordinates selection on the values of DOA and TOA.	65
4.5	Operational flow of the proposed joint propagation time delays and DOAs estimation using 2-D MP algorithms.	78
4.6	Comparison of computational complexity of the initial transformation and SVD of various 2-D MP algorithms.	83
4.7	Comparison of computational complexity of the SVD between 1-D UMP-Ex and 2-D UMP algorithms.	84

List of Figures

5.1	Deployment of BSs for wireless positioning based on TDOA and DOA.	86
5.2	STD of DOA estimates versus γ_{sc} (left) and DOA of the received signal (right).	105
5.3	STD of DOA estimates versus number of antennas (left) and system BW (right).	105
5.4	STD of time delay and positioning using TDOA error variances versus γ_{sc}.	109
5.5	STD of time delay and positioning using TDOA error variances versus system BW.	109
5.6	STD of time delay and positioning using TDOA error variances versus M.	110
5.7	STD of the first path time delay and relative DOA estimates versus the SNR.	114
5.8	STD of the first path time delay and relative DOA estimates versus system BW.	115
5.9	STD of the first path time delay and relative DOA estimates versus a number of antennas.	115
6.1	The RF Vector Signal Generator (Agilent MXG N5182A) and the Signal Analyzer (Agilent EXA N9010A).	119
6.2	The SDR modules USRP2 from Ettus and NI USRP-2921 from National Instruments.	120
6.3	The Network Analyzer (Agilent ENA E5071C).	121
6.4	The Wideband Time Domain Transmission (Agilent DCA 86100A).	122
6.5	The environment of RSS technique investigation based on the path loss model.	123
6.6	The estimated distances using the path loss model versus the actual distances.	123
6.7	System setup of fingerprinting investigation.	125
6.8	The performance of all 1-D MP algorithms at 20 MHz BW versus the pencil value.	126
6.9	Comparison of estimation accuracy of various 802.11ac BWs using 1-D UMP algorithm versus SNR.	127
6.10	The RMSE and complexity versus number of snapshots with SNR of 20 dB, and BW of 20 MHz.	127
6.11	The measured propagation time delays of path 1 and path 2 using the TDT.	129
6.12	The measurement system using Vector Signal Generator and Signal Analyzer.	130
6.13	Comparison of estimation accuracy and stability of various MP algorithms using single snapshot with Tx power of -20 dBm and IEEE 802.11a parameters.	130
6.14	Comparison of estimation accuracy and stability of various MP algorithms using single snapshot with Tx power of -20 dBm and IEEE 802.11n parameters.	131
6.15	Comparison of estimation accuracy and stability of UMP and UMP-Ex algorithms using 1 and 2 snapshots with Tx power of -30 dBm.	131
6.16	Frequency response measurement system using ENA with a reference signal.	133

List of Figures   xvii

6.17 The measured propagation time delays of path 1 and path 2 using the TDT, where Y-axis represents the derivative of output voltage of port 2. . 133
6.18 Comparison of estimation accuracy and stability of various 802.11ac BWs using single snapshot UMP-Ex algorithm. 134
6.19 System environment in both scenarios of LOS and NLOS wireless channels between the transmitter and the receiver with ULA. 135
6.20 Inside view of measurement locations in case of LOS wireless channel. . 135
6.21 The measurement system using ENA and the designed antenna array. . . . 136
6.22 Comparison of time delay estimation accuracy and stability of various 1-D MP algorithms using 40 MHz BW and 3 antenna elements in a LOS environment. 137
6.23 Comparison of estimation accuracy and stability of various 802.11ac BWs and possible frequency diversities using UMP-Ex algorithm and single antenna. 137
6.24 Comparison of time delay estimation accuracy of 1-D UMP-Ex using various 802.11ac BWs and different orders of ULA in a LOS environment. . 139
6.25 Comparison of time delay estimation accuracy of 1-D UMP-Ex using various 802.11ac BWs and different orders of ULA a LOS environment. . . . 139
6.26 Inside view of measurement locations in case of NLOS wireless channel. 140
6.27 Comparison of time delay estimation accuracy and stability of various 1-D MP algorithms using 40 MHz BW and 3 antenna elements in a NLOS environment. 140
6.28 Comparison of time delay estimation accuracy and stability of various 1-D MP algorithms using 160 MHz BW and 8 antenna elements in a NLOS environment. 141
6.29 Comparison of time delay estimation accuracy of 1-D UMP-Ex using various 802.11ac BWs and antenna array orders 1 to 8 in a NLOS environment. 142
6.30 Comparison of time delay and DOA estimation accuracy of various 2-D MP algorithms using 40 MHz BW and 3 antenna elements in a LOS environment. 143
6.31 Comparison of time delay and DOA estimation accuracy of 2-D UMP using various 802.11ac BWs and a number of antenna elements (3, 4, 6, or 8) in a LOS environment. 143
6.32 Comparison of time delay and DOA estimation accuracy of various 2-D MP algorithms using 40 MHz BW and 3 antenna elements in a NLOS environment. 144
6.33 Comparison of time delay and DOA estimation accuracy of various 2-D MP algorithms using 80 MHz BW and 8 antenna elements in a NLOS environment. 144
6.34 Comparison of time delay and DOA estimation accuracy of 2-D UMP using various 802.11ac BWs and a number of antenna elements (3, 4, 6, or 8) in a NLOS environment. 146
6.35 The lab and corridor used as test environment. 146
6.36 The locations of base stations and MU presented the relative angles. . . . 148
6.37 Comparison of XY coordinates estimation accuracy of various 802.11ac BWs and MU position estimators, where $M = 1$. 149

List of Figures

6.38 The STD of X coordinate estimates in cm. 149
6.39 The STD of Y coordinate estimates in cm. 150
6.40 Comparison of XY coordinates estimation accuracy of various oder of antenna arrays and 802.11ac BWs using W-LS estimator. 150
6.41 Localization results using 40 MHz BW and 3 antennas for each BS. . . . 151
6.42 CDFs of the position error e_z in cm for various number of snapshots based on the spatial diversity using 40 MHz BW. 152
6.43 Comparison of XY coordinates estimation accuracy of various 802.11ac BWs and MU position estimators while $M = 3$, based on the TDOA observations. 153
6.44 Comparison of XY coordinates estimation accuracy based on the TDOA observations for different number of array elements and 20 MHz BW. . . 153
6.45 The performance of W-LS estimator based on TDOA observations versus system BWs and antenna array orders. 154
6.46 Comparison of XY coordinates estimation accuracy based on DOA observations of various 802.11ac BWs and MU position estimators while $M = 4$. 155
6.47 Comparison of XY coordinates estimation accuracy based on DOA observations for different number of array elements and 40 MHz BW. 155
6.48 The distribution of the position error e_z of various estimators based on DOA observations (BW = 20 MHz, M = 4). 156
6.49 The distribution of the position error e_z of various estimators based on DOA observations (BW = 40 MHz, M = 4). 156
6.50 The distribution of the position error e_z of various estimators based on DOA observations (BW = 80 MHz, M = 4). 157
6.51 Comparison between the RMSE of W-LS estimator using TDOA observations, and W-SLS estimator using DOA observations. 159
6.52 The performance of hybrid DAC estimator versus number of antennas and system BWs. 159
6.53 The performance of the weighted TDOA and DOA estimator versus number of antennas and system BWs. 160
6.54 The performance of the proposed hybrid TDOA and DOA estimator versus number of antennas and system BWs. 160
6.55 The distribution of the position error e_z of various hybrid estimators based on TDOA and DOA observations. 161
6.56 Localization results using (BW = 20 MHz, $M = 2$) and (BW = 80 MHz, $M = 3$) 162
6.57 Performance comparison of 1-D UMP-Ex versus 2-D UMP. 163
6.58 CDFs of the position error e_z while $M = 2$. 164
6.59 CDFs of the position error e_z while $M = 3$. 164
6.60 CDFs of the position error e_z while $M = 4$. 165
6.61 Localization results using 40 MHz BW and 4 antennas for each BS using 1-D UMP-Ex. 165
6.62 Localization results using 40 MHz BW and 4 antennas for each BS using 2-D UMP. 167

List of Figures

7.1	The test environment.	172
7.2	The estimated time delays and the relative amplitudes of typical DDP and UDP channel profiles using 160 MHz BW.	173
7.3	Validity of modeling τ_{MED} by normal distribution.	175
7.4	Validity of modeling P_{loss} by normal distribution.	177
7.5	Validity of modeling κ by log-normal distribution.	178
7.6	Probability of UDP identification using the proposed methods and 802.11ac signals.	180
7.7	Another test environment (the stairs of building 2 and 3 in the university of Magdeburg).	182
7.8	The real environment between floor 3 and floor 2 of building 2 and 3 in the university of Magdeburg.	183
7.9	Probability of UDP identification using the proposed methods and 802.11ac signals, where the receiver is located under two floors.	183

List of Tables

3.1	The lengths of LTS and LTF regarding the 20 MHz BW of 802.11a and all BWs of 802.11ac.	44
3.2	The comparison between the principle of MP algorithm and that of the covariance matrix based algorithms.	46
3.3	The distribution of samples inside both possible generations of the extended matrix using the exchange matrix.	49
3.4	Computational Complexity of the Initial Transformation and SVD.	57
3.5	CFR vector length and pencil parameter value regarding the 20 MHz BW of 802.11a and all BWs of 802.11ac.	57
4.1	The selected values of pencil parameter K with respect to the number of antennas.	82
4.2	The selected values of pencil parameter P versus the CFR length.	83
4.3	The complexity ratio of 2-D UMP to the other 2-D matrix pencil algorithms.	83
5.1	List of estimators of MU position estimation.	100
6.1	The measured parameters of the path loss model.	122
6.2	The recorded RSS at radio map points with their coordinates.	125
6.3	Comparison between 1-D MP algorithms with spatial diversity and 2-D MP algorithms.	166
7.1	The mean and standard deviation of the normal PDFs for τ_{MED} and P_{loss}, and the log-normal PDF for $\ln(\kappa)$ using 802.11ac parameters.	180

Mathematical Notation

Symbol	Description
$\{.\}^*$	Conjugate operator
$\{.\}^T$	Transpose operator
$\{.\}^H$	Hermitian operator
$\otimes$	Kronecker product
$\odot$	Khatri-Rao product
$\nabla \mathbf{f}(\mathbf{z})$	Gradient of $\mathbf{f}(\mathbf{z})$ with respect to $\mathbf{z}$
$\dfrac{\partial f(\mathbf{z})}{\partial x}$	Differential of $f(\mathbf{z})$ with respect to x
$I!$	Factorial of I
$\mathbf{Y}^\dagger$	Moore-Penrose pseudo-inverse of matrix $\mathbf{Y}$
$\arg()$	Phase angle in radian
$\mathrm{avg}()$	Mean of the vector
$\mathrm{cov}()$	Covariance matrix
$diag\{\}$	Diagonal of a matrix
$E\{.\}$	Mathematical expectation
$\min\limits_{\mathbf{z}}$	Minimization over $\mathbf{z}$
$\mathrm{Im}\{\}$	Imaginary part of a given matrix
$\mathrm{Re}\{\}$	Real part of a given matrix
$\mathrm{range}\{\mathbf{Y}\}$	Range of matrix $\mathbf{Y}$ is the span of columns of $\mathbf{Y}$
$\mathrm{rank}\{\}$	Rank of a matrix
$\mathrm{vec}()$	Vector operation of a matrix (stacking each column of a matrix one under another)

List of Symbols

$h(t)$	Channel impulse response variable	14
L	Number of effective paths of multipath channel	14
α_l	Channel gain of the lth path	14
τ_l	Propagation time delay of the lth path	14
θ_l	Direction of arrival of the lth path	14
$a_m(\theta_l)$	Array response of the mth antenna to the lth path	14
$\delta(t)$	Dirac function	14
H	Channel frequency response variable	15
c	Speed of light	14
ρ	Distance between antenna elements in the antenna array	14
$x(t)$	Time domain transmitted signal	15
$r(t)$	Time Domain received signal	15
ω_c	Phase velocity of the carrier frequency	15
N_{FFT}	FFT length	15
Δf	Frequency spacing between sub-carriers	15
T	Sampling Interval	15
d	Distance between antenna pairs of communication link	17
ξ_d	Distance measurement error (DME)	19
τ_{RMS}	RMS delay spread	19
τ_{MED}	Mean excess time delay	19
ϕ	Phase of a received carrier signal	21
τ_m	Time error due to the rich multipath channel	23
τ_{UDP}	Time error due to the blockage of the direct path	23
τ_{pd}	Time error due to the induced propagation time delay	23
λ	Wavelength	24
P_t	Transmitted power	24
P_r	Received power	24
G_t	Transmitter gain	24
G_r	Receiver gain	24
χ_σ	Zero mean log-normally distributed random variable	25
n	Mean path loss exponent of pathloss model	25
σ	Standard deviation of pathloss model	25
I	Number of base stations	25
$\pm\Delta\theta$	Uncertainty of DOA estimation	28
$\mathbf{z}$	XY coordinates vector $\mathbf{z} = [x,y]^T$	32
d_{21}	Different distance between d_2 and d_1	34
$X_{i,k}$	Frequency domain transmitted symbol at subcarrier k during OFDM symbol i	38
N_u	Number of useful subcarriers	38
T_g	Guard interval duration	38
T_u	The useful part of the OFDM symbol	38
T_s	The total length of the OFDM symbol	38
$n(t)$	Time domain noise variable	38
ω_l^D	Phase velocity of Doppler shift of the lth path	38

$\Lambda(n)$	Cross-correlation output between the received samples and the training sequence	39
$\Upsilon(n)$	Moving sum of the received signal power	39
$R_{i,k}$	Frequency-domain received symbol of kth subcarrier during ith OFDM symbol	39
$w_{i,k}$	AWGN at the kth subcarrier during the ith OFDM symbol	39
f_l^D	Doppler spread of the lth path	39
f_c	Carrier frequency	39
N_p	Number of pilots	40
z_l	Multipath channel pole of lth path (frequency dimension)	40
v_l	Phase of multipath channel pole of lth path (frequency dimension) $v_l = 2\pi \Delta f \tau_l$	40
N_{dc}	Number of zero values at dc in the middle of training sequence	44
N	Number of pilots plus number of zero values at dc ($N = N_p + N_{dc}$)	44
$\mathbf{H}_{N\times 1}$	Channel frequency response vector	45
$\mathbf{Y}$	Hankel matrix	45
P	Pencil parameter value for frequency dimension	46
$\mathbf{Y}_1$	Submatrix of matrix pencil by deleting the last row	46
$\mathbf{Y}_2$	Submatrix of matrix pencil by deleting the first row	47
$\mathbf{Z}_d$	Diagonal matrix of the multipath channel poles	47
$\mathbf{A}$	Diagonal matrix of channel gains	47
$\mathbf{I}$	Identity matrix	47
$\mathbf{Y}_{ex}$	Extended matrix	47
Π	Exchange matrix	48
$\mathbf{U}$	Unitary matrix of left eigenvectors	48
$\mathbf{V}$	Unitary matrix of right eigenvectors	48
Σ	Diagonal matrix with the singular values	48
Σ_s	Diagonal matrix with the singular values of the signal subspace	48
Σ_n	Diagonal matrix with the singular values of the noise subspace	48
η_k	singular value of column k	48
σ_l^2	noise variance of lth path	48
$\mathbf{U}_s$	Submatrix of $\mathbf{U}$ that span the signal subspace	49
$\mathbf{V}_s$	Submatrix of $\mathbf{V}$ that span the signal subspace	49
$\mathbf{U}_n$	Submatrix of $\mathbf{U}$ that span the noise subspace	49
$\mathbf{V}_n$	Submatrix of $\mathbf{V}$ that span the noise subspace	49
$\prod$	Product	48
$\sum$	Summation	48
$\mathbf{Y}_s$	Hankel matrix of the signal subspace	48
$\mathbf{Y}_n$	Hankel matrix of the noise subspace	48
$\mathbf{z}$	Multipath channel poles vector (frequency dimension)	49
$\mathbf{A}_0$	Channel gains vector	50
$\mathbf{Y}_{\text{Re}}$	Real matrix	50
$\mathbf{Q}_q$	Unitary matrix whose columns are conjugate symmetric and have a sparse structure	50
$\mathbf{0}$	zero row vector	50
$\mathbf{J}_1$	Selection matrix in 1-D UMP	50
$\mathbf{J}_2$	Selection matrix in 1-D UMP	50
$\mathbf{K}_{\text{Re}}$	Selection matrix due to using the UMT	51
$\mathbf{K}_{\text{Im}}$	Selection matrix due to using the UMT	51
$\mathbf{F}_C$	DFT matrix of dimensions $C \times C$	52

List of Tables

$\mathbf{f}_m$	The mth row of DFT matrix, DFT beam steered at spatial frequency $\nu = m \times 2\pi/C$	52
$\mathbf{Y}_F$	The matrix $\mathbf{Y}$ multiplied by DFT matrix	53
$\mathbf{B}$	Real-valued beam-space array manifold matrix	53
$b_m(\nu_l)$	The (mth, lth) element of the DFT beam-space manifold $\mathbf{B}$	53
$\mathbf{\Gamma}_1$	Selection matrix in the 1-D BMP algorithm	53
$\mathbf{\Gamma}_2$	Selection matrix in the 1-D BMP algorithm	53
$\mathbf{T}$	Non-singular matrix	54
$\mathbf{Y}_E$	Enhanced matrix for multiple snapshot principle	55
q	Number of multiple snapshots	55
$\mathbf{a}(\theta_l)$	Steering vector or array manyfold	62
$R_{m,k}$	Frequency-domain received symbol by the mth antenna and at the kth subcarrier	62
$H_{m,k}$	CFR sample of the mth antenna and the kth subcarrier	62
$\mathbf{H}_{M \times N}$	Space-channel frequency response	63
x_l	Multipath channel pole of lth path (space dimension)	63
μ_l	Phase of multipath channel pole of lth path (space dimension) $\mu_l = 2\pi f_c \rho \sin\theta_l / c$	63
$\mathbf{Y}_m$	Hankel matrix of antenna m	66
$\mathbf{Y}_e$	Hankel block matrix	66
K	Pencil parameter value for space dimension	66
$\mathbf{X}_d$	Diagonal matrix including the eigenvalues of space dimension	67
$\mathbf{P}$	Shuffling matrix	67
$\mathbf{U}_{sp}$	Submatrix of $\mathbf{U}$ that span the signal subspace and shuffled by the shuffled matrix $\mathbf{P}$	69
$J_s(i,j)$	Correlation maximization criterion	70
$\mathbf{W}$	Eigenvectors matrix of eigenvalue problem	71
$\mathbf{W}'$	Modified eigenvectors matrix of eigenvalue problem	71
$\mathbf{J}_{v1}$	Selection matrix in 2-D UMP for frequency dimension	72
$\mathbf{J}_{v2}$	Selection matrix in 2-D UMP for frequency dimension	72
$\mathbf{J}_{u1}$	Selection matrix in 2-D UMP for space dimension	72
$\mathbf{J}_{u2}$	Selection matrix in 2-D UMP for space dimension	72
$\mathbf{P}'$	Modified shuffling matrix due to using the UMT	73
$\mathbf{K}_{\mathrm{Re}1}$	Selection matrix due to using the UMT for space dimension	73
$\mathbf{K}_{\mathrm{Im}1}$	Selection matrix due to using the UMT for space dimension	73
$\mathbf{K}_{\mathrm{Re}2}$	Selection matrix due to using the UMT for frequency dimension	73
$\mathbf{K}_{\mathrm{Im}2}$	Selection matrix due to using the UMT for frequency dimension	73
η_d	Diagonal matrix of eigenvalues	74
$\mathbf{\Gamma}_{\mu 1}$	Selection matrix in 2-D BMP for space dimension	77
$\mathbf{\Gamma}_{\mu 2}$	Selection matrix in 2-D BMP for space dimension	77
$\mathbf{\Gamma}_{v 1}$	Selection matrix in 2-D BMP for frequency dimension	77
$\mathbf{\Gamma}_{v 2}$	Selection matrix in 2-D BMP for frequency dimension	77
ϱ	The complexity ration of 1-D MP to 2-D MP algorithms	82
Δt_{ij}	TDOA between receivers i and j	85
$\mathbf{G}$	Designed Matrix	86
$\mathbf{h}$	Designed Matrix	86
$\mathbf{Q}_{TDOA}$	Covariance matrix of TDOA observations in distance form	86
$\hat{\mathbf{z}}_{ML}$	The ML position estimate of MU coordinates	87
σ^2_{DOA}	Variance of DOA observations	92
$\mathbf{Q}_{DOA}$	Variance matrix of DOA observations	92

$\sigma^2_{r,TDOA}$ Variance of TDOA observations in distance form 95
$\sigma^2_{r,DOA}$ Variance of DOA observations in distance form 95
κ_{TDOA} Weight for TDOA observations 95
κ_{DOA} Weight for DOA observations 95
$\sigma^2_{\tau,CRB}$ CRB of time delay error variance 99
$\sigma^2_{\theta,CRB}$ CRB of DOA error variance 99
β Channel Gain $\beta = \alpha e^{-j2\pi f_c \tau}$ 100
$\mathbf{Q}_n$ Covariance of Gaussian noise $\mathbf{Q}_n = \sigma^2_n \mathbf{I}$ 101
$\mathbf{F}_{ij}$ The $(i,j)th$ of the Fisher information matrix (FIM) 101
γ_{sc} SNR per subcarrier 103
$\sigma^2_{\theta,CRB,OFDM}$ CRB of DOA error variance of OFDM signals 103
$\sigma^2_{\tau,CRB,OFDM}$ CRB of time delay error variance of OFDM signals 104
$\mathbf{F}_{OFDM,SD}$ FIM of the OFDM signal with spatial diversity 107
$\sigma^2_{\tau,CRB,OFDM,SD}$ CRB of time delay estimation for OFDM signals with spatial diversity 107
$\sigma^2_{r,TOA}$ Variance of time delay observations in distance form 107
$\mathbf{U}(\boldsymbol{\theta},\boldsymbol{\tau})$ Space-frequency response matrix for L paths, N subcarriers, and M antennas 110
$\sigma^2_{\eta,CRB,ULA-OFDM}$ CRB of joint time delay and DOA estimation for ULA-OFDM systems . 113
$e_\mathbf{z}$ The distance between the estimated position and the actual position of the MU 147
$p(x/S_i)$ PDF of variable x given hypothesis S_i 174
P_{tot} The total power of the received signal 174
P_{loss} The power loss of the transmitted signal 174
κ Hybrid time-power parameter 176

List of Abbreviations

1-D One-dimensional

2-D Two-dimensional

3-D Three-dimensional

A-DAC-LI Adapted DAC based on lines intersection

A/D Analog to digital converter

AD Anderson-Darling

AGC Automatic gain control

AOA Angle of arrival

AP Access point

AWGN Additive white Gaussian noise

BMP Beam-space Matrix Pencil

BS Base station

BW Bandwidth

CDF Cumulative distribution function

CDMA Code division multiple access

CFR Channel frequency response

CIR Channel impulse response

COTS Commercial of the shelf

CP Cyclic prefix

CRB Cramer-Rao bound

CS Compressive sensing

DAC Divide-and-Conquer

DAC-LI DAC based on lines intersection

DAC-LS DAC-Least Square

D/A Digital to analog converter

DCA Digital Communication Analyzer

DDP Detected direct path

DFT Discrete Fourier Transform

DME Distance measurement error

DOA Direction of arrival

DP Direct path

DSP Digital Signal Processor

DSSS Direct sequence spread spectrum

ECDF Empirical cumulative distribution function

ESPRIT Estimation of Signal Parameters via Rotational Invariance Technique

EVD Eigenvalue decomposition

FDP First detected path

FFT Fast Fourier transform

FIM Fisher information matrix

FPGA Field Programmable Gate Array

GPS Global Positioning System

H-DAC Hybrid DAC based on TDOA and DOA

HT High-throughput

HT-LTF High-throughput Long Training Field

Hybrid-LS Hybrid TDOA and DOA

Hybrid-W Weighted TDOA and DOA

ICZT Inverse chirp Z transform

IFFT Inverse fast Fourier transform

ILS Iterative least square

IR-UWB Impulse radio UWB

ISI Intersymbol interference

ITC Information Theoretic Criteria

L-LTF Low-throughput long training field

L-STF Low-throughput short training field

List of Tables

LBS Location Based Services

LI Lines Intersection

Loran LOng RAnge Navigation

LOS Line of sight

LS Least Square

LTS Long training symbol

MBMP Multiple Invariance BMP

MDL Minimum Descriptive Length

MIMO Multiple input multiple output

ML Maximum likelihood

MP Matrix pencil

MU Mobile unit

MUSIC Multiple Signal Classification

NDP Null data packet

NIC Network Interface card

NLOS Non-line of sight

ns Nanosecond

NUDP Natural UDP

OFDM Orthogonal frequency-division multiplexing

PDF Probability density function

PLCP Physical layer convergence procedure

PN Pseudo noise

POA Phase of Arrival

PPDU PLCP protocol data unit

ps Picoseconds

PSDU PLCP service data unit

RP Reference point

RFID Radio frequency identification

RMS Root mean square

RMSE Root mean square error

RSS Received signal strength

RT Ray tracing

S-CFR Space channel frequency response

SBMP Single Invariance BMP

SCPI Standard Commands for Programmable Instruments

SDR Software defined radio

SLS Stansfield Least Square

SNR Signal to noise ratio

SP Strongest path

STD Standard deviation

STS Short training symbol

SUDP Shadowed UDP

SV Singular value

SVD Singular Value Decomposition

TDOA Time difference of arrival

TDT Time domain transmission

TOA Time of arrival

UDP Undetected direct path

ULA Uniform linear array

UMP Unitary Matrix Pencil

UMT Unitary matrix transformation

URA Uniform rectangular array

USRP Universal Software Radio Peripheral

UWB Ultra-wide band

VHT Very high-throughput

VHT-LTF Very high-throughput Long Training Field

List of Tables

W-ILS   Weighted Iterative Least Square

W-LS   Weighted Least Square

W-SLS   Weighted Stansfield Least Square

WLAN   Wireless Local Area Network

CHAPTER 1

Introduction to Wireless Indoor Positioning

Wireless indoor positioning can be defined as the estimation of the mobile unit (MU) coordinates in the required area. This can be obtained by periodically transmitting properly designed signals from the MU and receiving them at a number of fixed base stations (BSs) with known position coordinates to the system. The MU coordinates are then estimated by processing the received signals. The characteristics of the received signals are used to localize the MU based on the interaction between the transmitted signals and the wireless channel. The BSs could be called access points (APs) or reference points (RPs).

Wireless positioning using radio signals has received considerable attention in the field of navigation and tracking. The best example of this difficult problem is the Global Positioning System (GPS) [2]. It is designed to find the position of GPS receiver in the open environments, where the GPS receiver and at least four GPS satellites are in line of sight (LOS). However, the existing GPS cannot provide accurate position estimation in urban and indoor environments due to shadowing and various sources of noise and interference. The Global Navigation Satellite System (GNSS) is another example for navigation based on satellites. A classical example of a terrestrial radio navigation system is Loran-C (LOng RAnge Navigation) [3]. The basic Loran-C system consists from a number of land-based transmitting stations, each separated by several hundred miles. One station from Loran stations is designed as a master station, and the other transmitters as secondary stations. In practice, the receiver simply observes the time differences between the received signals of three Loran stations or more, and then converts the measured time differences to more commonly-used coordinates, such as a latitude and longitude, using special charts [3]. In fact, wireless indoor positioning is very different from those traditional systems in many aspects such as the operating environments, system requirements, and the required applications including the required accuracy [4]. Therefore, there is a need for new positioning systems that are specifically designed for indoor applications and can overcome the challenges of indoor environments. As a result, the wireless indoor positioning is emerging as a new important research area.

1.1 Basic Observations

This work is mainly focused on the wireless positioning using the characteristics of radio signals. Therefore, the position estimation of MU will be derived from the measured radio signals. There are three basic properties that can be used for MU position estimation from analysis of specific physical characteristics of received radio signals, which are related to the relative position of the MU with respect to the fixed BSs. Those properties can be used in different ways; however, they can be classified into three categories. The first category is based on assigning each point of space a unique set of signal strengths received from the properly distributed BSs. The received signal strength (RSS) technique is the simplest as regarding signal processing complexity. The second category uses the propagation time of the signals as an estimate for the distance between a BS and the MU. The time-based techniques can be grouped in time of arrival (TOA) and time difference of arrival (TDOA) techniques. In TOA techniques, the MU and BSs should be synchronized while in TDOA techniques clock synchronization is only required among the fixed BSs. Usually, they are connected to a wired backbone, which simplifies the synchronization between them. The third category is based on the directional property of antenna arrays for detecting a direction of arrival (DOA). The related signal processing is similar to that used in digital beam forming. In fact, the characteristics of radio signals, namely RSS, DOA, TOA, and TDOA can be used alone or in combinations.

The wireless indoor positioning systems based on radio signals can be actually grouped into two categories: distance-based techniques and direction-based techniques [5]. The distance-based techniques depend on the estimation of distances between the MU and a number of fixed BSs. The required observation could be the RSS, the phase of the carrier signal, or the TOA of the received signal. The DOA is the common observation used in the direction-based systems.

Wireless indoor positioning systems can also be grouped into two categories: multilateral or network-based architecture and unilateral or mobile-based architecture [6], [5]. For both cases, a number of BSs with known coordinates are used to estimate the position of MU based on the above characteristics of radio signals as follows:

- *Unilateral system:* In a unilateral system, the MU receives transmissions from a number of BSs, extracts the required observations from the received radio signals, calculates its position coordinates, and then displays the position on its screen. As a consequence, the MU is a complex device. GPS is a unilateral system. The multiple transmitting satellites send their clock reading at the instant of epoch transmission and their accurate position information; GPS requires one-way communication between satellites and GPS receiver. A classical example of TDOA unilateral system is Loran-C [3].

- *Multilateral system:* In a multilateral system, a number of BSs receive the transmitted signal from the MU, and then extract the required observations. After that, the BSs report the estimated observations to a central unit to calculate the MU coordinates using the MU positioning algorithms. The MU coordinates can be forwarded to the MU from the central unit. Hence, the MU can be implemented much simpler than in a unilateral system. Cellular positioning is a multilateral system, where several BSs report the time of reception from the MU to the central unit to estimate the MU position.

From the above, the normal characteristics of radio signals are used to estimate the MU position. In addition, there are three principles that are different from the above principles:

- *Fingerprinting:* The basic principle is to compare the measured observations such as the RSS with the database, which has been already created in an advance in the initial training phase for the required area as in [7], [8].

- *Ray Tracing:* The basic principle is to represent the electromagnetic waves as rays and to produce deterministic channel models that operate by processing user-defined environments as in [9], [10], [11].

- *Proximity:* The basic principle is to detect the MU in the range of a fixed BS, so the MU is known to be within area around a known position as in [12].

Although the RSS can be obtained directly from the Network Interface Cards (NIC) that are available in most wireless devices, and it can be applied using wireless sniffer software tools, the accuracy of RSS techniques decreases with greater distances due to the fact that the free space attenuation increases with the logarithm of the distance. The RSS may change by 10 dB and more due to multipath effects and the orientation of wireless devices [13]. It is sensitive to the changes in the indoor environments. In addition, the pathloss is significantly affected by the materials of walls and floors, layout of each floor, number of floors, and the location of obstructing objects. As a result, it is difficult to find a general model for all environments. The time-based techniques (TOA and TDOA) are not suffered from these problems. However, they are affected by the multipath effects, the signal-to-noise ratio (SNR), and the system bandwidth. In addition, they require a synchronized network. The major drawback of wireless positioning based on TOA is that the MU and the BS should be synchronized with high accuracy. It means that the MU requires highly stable oscillator and robust hardware for time stamping [13]. Our goal is that the MU should be a very simple device, where the complexity could be added only to the network side. Therefore, the TDOA based techniques are preferred for a high-resolution wireless positioning. In Chapter 2, the introduced characteristics of received radio signals with their advantages and disadvantages will be presented in more details.

1.2 Wireless Indoor Positioning Applications

Wireless indoor positioning is a popular research topic, its services will be widely used in the future through a number of applications. The wireless positioning applications are often referred to the Location-Based Services (LBS). The applications of wireless indoor positioning systems are vast, and can be broadly grouped into some categories such as health care, commercial, public safety, and military applications as shown in Fig. 1.1 [14].

In the health care applications, the residential and nursing homes have an increasing demand for indoor positioning systems to track people with special needs, the elderly, and children who are away from visual supervision, to guide the blind, to locate instruments and other equipments inside hospitals, and to locate surgical equipments in an operating room [15].

The commercial applications include an inventory tracking in a warehouse, supply-chain management or workflow optimization, an interactive tour guide for museums, location-sensitive web-browsing, and so on [16]. Recently, the pedestrian foot traffic in shopping malls can be tracked by determining the positions of shoppers using the control channel transmissions of cell phones [17].

In the public safety and military applications, precise indoor positioning systems are needed to assist policeman, fire fighters, and soldiers to complete their missions inside buildings [18]. Accurate indoor positioning is also an important part of various mobile robotics applications [19]. This will be a great help to the automation field such as a mobile robot self-locating and navigation, and intelligent building [20].

More recently, wireless positioning has found applications in location-based handoff in wireless networks, location-based ad-hoc network routing, and location-based authentication and security [21]. Positioning and tracking offer also an effective solution for context-aware applications that not only respond, but anticipate user needs [22]. Wide range of applications can be developed using sensor networks and radio-frequency identification (RFID) technologies such as locating unwanted chemical, biological, or radioactive material using sensor networks, and tracking specific items using RFID tags [15]. Using Wireless Local Area Network (WLAN) is increasing in industrial environments, which opens other possibilities besides the communication such as the positioning of people, end-devices, machine parts, and so on [23].

From the applications listed above, clearly the required accuracy of wireless indoor positioning depends on the application. It varies from a few millimeters in surgical navigation, tumor detection, or sensitive nuclear material tracking to a few meters for Tractor inventory tracking in a warehouse [18].

1.3 The Main Challenges of Wireless Indoor Radio Propagation Channel

In a wireless communication channel, the transmitted signal may suffer many reflections and dispersion due to obstacles when it propagates. Such obstacles may be stationary or moving with time. Hence, multipath reception is a characteristic of the wireless channel. At the receiver side, several versions of the same signal are received with different amplitudes, phases, angle of arrivals, and time delays due to different paths.

If we look to Fig. 1.2 for a while, the wireless indoor radio propagation channel properties can be extracted. The wireless indoor channel can be characterized as a site-specific channel, where every environment has its own characteristics, and it is impossible to find a general model for the indoor environments from the point of positioning. It is a severe multipath channel as a result of harsh environments. Therefore, a wireless indoor channel is always unpredictable. The probability of LOS between a transmitter and a receiver is rare, where the transmitter is almost of the time in non-line of sight (NLOS), or maybe in a blockage of the direct path, caused by the presence of walls, humans, and other rigid objects. In addition, indoor environments have been built from different media,

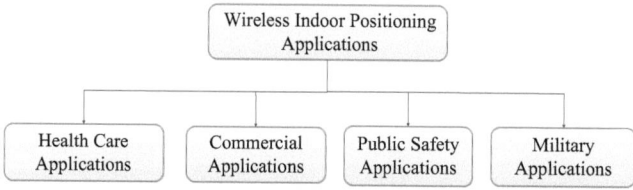

Figure 1.1: The wireless indoor positioning applications.

which means that there are different propagation time delays. Last but not the least is bandwidth limitations. System bandwidth (BW) is the core of wireless indoor positioning systems that are based on time observations; it is a very important parameter that effects on the performance. Using large bandwidths involves high resolution of paths distinction. From the above challenges, including also the system complexity, and security, we can say that accurate indoor tracking is very challenging for the scientific community. The block diagram of Fig. 1.3 summarizes the properties of wireless indoor propagation channel.

1.4 Brief Literature Overview

Many of the studies have been proposed to estimate the TOA based on the statistical principles such as Estimation of Signal Parameters via Rotational Invariance Techniques (ESPRIT) [24], Multiple Signal Classification (MUSIC) [4], [25], [26], and Root MUSIC [20], which are unattractive for online estimation. The number of collected channel estimates should be at least as the number of multipaths. The channel estimates should be taken during a time interval larger than the coherence time of the channel to satisfy the full rank condition [27]. The principle of ESPRIT has also been used to estimate the DOA such as in [28], [29], the same in [30] for Root MUSIC. The problem of joint estimation of angles and relative time delays of multipath signals has been addressed for narrowband signals using ESPRIT in [27], [31]. The problem of joint estimation of time delays and relative two-dimensional (2-D) DOA of multipath signals has been addressed for narrowband signals using ESPRIT in [32]. The problem of hybrid TDOA and DOA has been investigated in [33] for Code division multiple access (CDMA) cellular systems.

Recently, a non-statistical algorithm called the matrix pencil (MP) has been used to estimate the DOA of narrowband signals using the uniform linear array (ULA) as in [34], [35]. The principle of the conventional MP algorithm was enhanced in [36] to estimate 2-D frequencies using the uniform rectangular array (URA). It was also enhanced to estimate the azimuth and elevation angles using the URA as in [37], [38]. The MP algorithm has been used for ultra-wide band (UWB) wireless sensor networks in [39], and for impulse radio UWB (IR-UWB) in [40]. A comparison between the conventional MP algorithm, and the statistical super resolution algorithms (ESPRIT and Root MUSIC) has been presented in [41]. The principle of MP algorithm is superior compared to that of ESPRIT, MUSIC, and Root MUSIC, as it will be presented in Section 3.4. Therefore, in this work, the principle of MP algorithm will be enhanced to estimate the required channel profile parameters.

The TDOA and DOA estimates are used to estimate the MU position. To do that using TDOA estimates, the intersection of hyperbolic curves defined by TDOA observations should be determined by solving a set of nonlinear equations. There are many algorithms to solve nonlinear equations, most of them are time-consuming and inconvenient for implementation. The iterative least square (ILS) estimator is presented in the literature using Taylor series expansion to linearize those nonlinear equations [42]. It starts with an initial guess for the MU position, and then calculates the position deviation for each iteration. The drawback of ILS estimator is that it is a computationally intensive method if the starting point is not close enough. Moreover, it has a convergence problem [43]. Divide-and-Conquer (DAC) method presented in [44] can achieve a good performance at high SNR. The principle of DAC method is to split the observations to

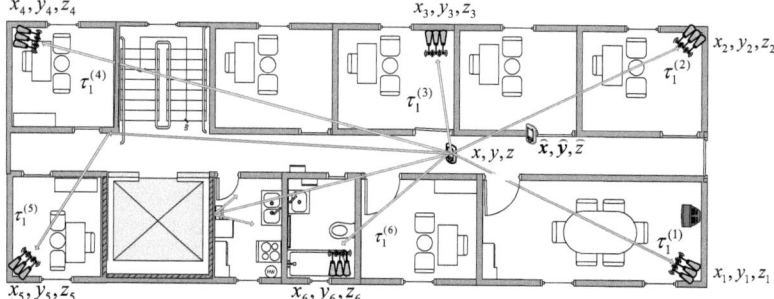

Figure 1.2: Multipath radio environment.

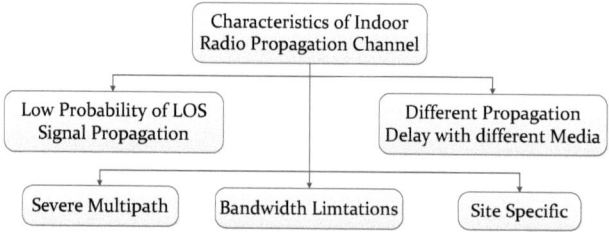

Figure 1.3: Characteristics of wireless indoor radio propagation channel.

small sets each having a size equal to the number of unknowns. The drawback of this method is that those sets should have a large Fisher information [45]. Chan proposed an estimator in [43] that assumes initially the MU coordinates and the reference distance to the reference BS are independent, and the MU has identical distances to all BSs, in another way, the MU is located in the far field from the system. The drawbacks of Chan estimator are that additional steps are necessary to compensate those assumptions to get the final position estimate, and the performance is very weak at low system conditions; it is based on the assumption that the noises in the TDOA measurements are small. For DOA estimates, the MU position is found by triangulation. The principle of DAC method based on the TDOA observations can be used based on the DOA observations [46]. To solve the problem of partitioning, a non-iterative closed form has been proposed based on the DOA measurements in [47], which is called a Stansfield estimator.

However, some of those studies employ narrowband signals; others are complex and based on statistical principles, or restricted to far-field assumptions. In addition, the multipath indoor radio channels are complex and cannot be analyzed with computer simulations with simple theoretical channel models. In fact, the performance should by analyzed based on the experimental measurements to get much more realistic results. Many related references in the literature will be surveyed in details and referred in later chapters where it is appropriate.

1.5 Objectives of the Dissertation

To design a wireless positioning system, there are basically two approaches [18], [15]. The first approach is to develop a new signaling system and a network infrastructure for the required positioning application. The system parameters and as a result the expected performance can be controlled by the designer. The size of MU can be designed to be a very small device. The reference BSs can also be distributed based on the required accuracy. The second approach is to use an existing wireless network infrastructure. Using this approach avoids the deployment of a new infrastructure. However, intelligent algorithms should be used to compensate the lack in system parameters.

In this work, the main objective is to develop a wireless indoor positioning system based on the existing wireless network infrastructure. The opportune signals of IEEE 802.11 standards can be used for an effective wireless indoor positioning. To build an accurate indoor positioning system, the time-based techniques should be used to estimate the MU position. In these techniques, system bandwidth is an important parameter that effects on the performance of the indoor positioning system. The bandwidths of the order of 10 MHz used in GPS are not sufficient for wireless indoor positioning. In fact, bandwidths of the order of several hundred megahertz should be used to provide a reasonable protection against extensive multipath environments in indoor areas [4].

IEEE 802.11ac is an emerging WLAN standard for the 5 GHz band in which multiple input multiple output (MIMO) and wider channel bandwidths are enhanced compared to 802.11n. To support wider channel bandwidths, 802.11ac defines its channelization for 20, 40, 80, and 160 MHz channels [48]. The largest channel bandwidth of 802.11a is 20 MHz, and of 802.11n is 40 MHz. Hence, due to the bandwidth limitation of IEEE 802.11 standards, super resolution algorithms are used for post processing to reduce bandwidth requirements. The emerging 802.11ac standard provides as a maximum 8×8 MIMO antenna configuration. It has been enhanced compared to 802.11n, which is 4×4 [49], and to 802.11a, which is 1×1 [50]. While the main advantage of MIMO is to enhance data throughput, it can also be used to estimate the DOA. In addition, it can be used as a kind of spatial diversity for TDOA estimation besides the frequency diversity coming from the orthogonal frequency-division multiplexing (OFDM). Therefore, a framework for wireless indoor positioning using recent wireless networks with MIMO-OFDM technology needs to be developed and investigated. The developed algorithms should use a very efficient computational methodology for real time implementation on a digital signal processor (DSP) chip.

The main objectives of this work could be broken down into the following majors:

- The problem of highly resolving the propagation time delays of multipath signals will be addressed for 2-D wireless indoor positioning. In this work, the problem of estimating the TDOA associated with signals in multipath communication channels will be investigated using a single transmitter and a number of receiving BSs. In wireless communication systems, the training sequences are used to estimate synchronization and channel parameters. Hence, the key element of our work is to use the preamble of the OFDM frame to measure the channel state for additional purposes to the demodulation of the data portion of the physical layer convergence procedure (PLCP) protocol data unit (PPDU), which contains the training fields [48], [49], [50]. It means that the frame format will be used without any modification.

- The diversity techniques such as time diversity, frequency diversity, and space diversity are widely utilized in wireless communication systems to improve the performance of the communication link [51]. In this work, the principle of using diversity techniques will be investigated to improve the performance of the positioning system; using diversity techniques leads to get the advantage of the random nature of the radio propagation channel by finding and combining uncorrelated paths [4].

- The problem of highly resolving the propagation time delays and the relative DOAs of multipath signals will be investigated using a single transmitter and a number of receiving BSs for 2-D wireless indoor positioning system based on the hybrid TDOA and DOA measurements.

- A closed-form solution to the estimation problem of MU coordinates should be developed. A non-iterative estimator should be developed to estimate the MU coordinates based on the TDOA observations. Furthermore, a non-iterative hybrid estimator should be developed based on the TDOA and DOA observations. The proposed estimators should solve the problem of the initial guess and that of partitioning of the other estimators.

- Estimating an accurate position in indoor areas involves many difficulties. One of those difficulties is the high probability of NLOS signal propagation. It causes undetected direct path (UDP) conditions, which represent a serious challenge to the design of an accurate wireless indoor positioning system. As an example for the critical locations in NLOS scenarios is that the MU is located behind a metallic chamber or an elevator, or there are many rigid walls between the MU and the

BS. Therefore, one of the major challenges for wireless indoor positioning is the identification of UDP channel profiles. Adding the channel obstruction knowledge improves the accuracy of the positioning system. The results can be discarded or rectified if there is a limited connectivity.

1.6 The Dissertation Outline and Contributions

The outline of the dissertation and the author's contributions are presented in the following. The main contributions of each chapter will be presented under its outline, although all measurements have been collected in Chapters 6 and 7.

Chapter 2: Fundamentals of Position Estimation Techniques

The fundamentals of position estimation techniques have been investigated. The advantages and disadvantages of using various radio signal characteristics are presented, which have been introduced in [52].

Chapter 3: System Model and Time Delay Estimation using 1-D Matrix Pencil Algorithms

The capability of using OFDM systems in wireless indoor positioning is investigated. The power of carrier frequency is zero in OFDM systems, hence, there is a discontinuity in the OFDM spectrum. The discontinuity of the estimated channel frequency response (CFR) at dc should be removed; consequently, the time delay estimation problem of OFDM signals in a multipath channel is fully equivalent to the DOA problem in antenna array processing. The preamble of the OFDM frame has some training fields; the best field has been selected to estimate the CFR, published in [53].

The main requirements of many applications using recent wireless networks are the high accuracy and low complexity. For high-resolution wireless positioning, the TDOA associated with signals in a multipath communication channel should be estimated. Recently, variants of MP algorithms have been presented to estimate the DOA of coherent or non-coherent narrowband signals using the ULA. These include the Unitary Matrix Pencil (UMP), the Single Invariance, and the Multiple Invariance Beam-space Matrix Pencil (BMP) algorithms, which are non-statistical algorithms and based on the real computations. In this chapter, the various one-dimensional (1-D) MP algorithms have been enhanced and implemented in different realizations to estimate the propagation time delays and the relative amplitudes from the estimated CFR using OFDM systems.

The accuracy, stability, and complexity of various 1-D MP algorithms are investigated using 802.11a and 802.11n system parameters, where one OFDM training symbol and 20 MHz bandwidth are used, published in [53]. The performance of various MP algorithms is also investigated using the emerging 802.11ac standard, and compared to the corresponding performance of 802.11n and 802.11a. The performance of using multiple OFDM training symbols as a kind of temporal diversity and wider channel bandwidths of 802.11ac are emphasized, published in [54]. The performance of using wideband orthogonal multi-carrier signals with spectral diversity is presented. The complexity of using the high BWs of 802.11ac has been treated and reduced based on the frequency diversity, published in [55].

The various MP algorithms can also be enhanced and applied in a new way to estimate the propagation time delays from the space channel frequency response (S-CFR) using multi-antenna multi-carrier systems (MIMO-OFDM systems) for high-resolution wireless positioning. Considerable improvement using the spatial diversity has been presented especially at low SNR and narrow BWs; it represents a robust technique versus multipath channel fading. In fact, the problem of arrays imperfection and orientation occurred in DOA can be mitigated, published in [56].

Chapter 4: Joint Time Delay and DOA Estimation using 2-D Matrix Pencil Algorithms

In this chapter, the problem of highly resolving the propagation time delays, the relative DOAs, and the relative amplitudes associated with signals in multipath communication channels for a high-resolution wireless positioning system has been addressed. The principle of how to use multi-antenna multi-carrier systems for wireless positioning is revealed. The recent subspace-based algorithms, represented by 2-D MP, 2-D UMP, and 2-D BMP algorithms, are applied in a new way to estimate these parameters simultaneously from the measured S-CFR using multiple antennas and wideband orthogonal multi-carrier signals.

To reduce the complexity of joint time delay and DOA estimation problem, the priori information of wireless positioning (our concern is to estimate the time delay and the relative DOA of the first path) is used, which mitigates the problem of repeated poles and hence reduces the complexity of calculating extra eigenvalue decomposition problems, published in [57]. The computational complexity of various 2-D MP algorithms has been derived precisely. The complexity of data matrix transformation, the singular value decomposition, the eigenvalues computation using QR and QZ algorithms, and the other necessary steps has been derived precisely. Results have been presented in [58]. It is known that to increase the DOA accuracy, the number of antennas in the antenna array should be increased, which increases the complexity of BS. However, the necessary number of antennas for accurate DOA estimation can be reduced to reduce the complexity by increasing the number of subcarriers (the bandwidth). In another way, a compensation for the lack in the system BW or in the array order has been presented to get accurate estimation for time delays and relative DOAs. It has been found that using multi-antenna multi-carrier principles can successfully enhance the dimensionality of the signal subspace for joint time delay and DOA estimation; it represents a robust technique versus multipath channel fading, published in [57].

From the previous, it is worth mentioning that in addition to spectral diversity coming from OFDM, two principles are presented to use the spatial diversity in the enhanced 1-D and 2-D MP algorithms in Chapters 3 and 4, respectively. After estimating the required observations, the next step is to estimate the MU coordinates using TDOA observations, or hybrid TDOA and DOA observations. The cost function which takes into account the accuracy of TDOA and DOA during the combination between the estimated coordinates of both of them should be investigated to take into account the amount of accuracy of each observation.

1.6. The Dissertation Outline and Contributions
Chapter 5: Mobile Unit Position Estimation Based on TDOA and DOA Measurements

First, the principle of some useful studies in the literature will be presented to show later the performance of the proposed estimators. Then, a non-iterative weighted least square (W-LS) estimator is presented to estimate the MU coordinates based on the TDOA estimates, published in [56]. Furthermore, a non-iterative hybrid W-LS estimator has been presented based on the TDOA and DOA estimates. Results have been presented in [58]. The proposed estimators should solve the problem of the initial guess and that of partitioning of the other estimators. They should also be robust against channel fading and low SNR.

The positioning accuracy is limited by the fundamental parameters of a wireless positioning system such as the number of antenna elements in the array, number of subcarriers including subcarrier spacing, SNR, the estimated DOAs with respect to the array, and many others. To measure the effect of those parameters on the positioning performance (the positioning error variance), the Cramer-Rao bounds (CRBs) of TDOA and DOA estimation error variances have been derived. To combine TDOA and DOA, the accuracy of each type is different due to the different nature of each principle. Both observations should be combined in an optimal way, therefore, the TDOA and DOA observations should be given appropriate weights.

Chapter 6: Performance Evaluation Based on Channel Measurements

In the first part, equipments description that were used in this work is presented. Then, the expected performance of using RSS techniques for wireless positioning has been confirmed. In the remaining part, the performance of the proposed algorithms in the previous chapters using 802.11a, 802.11n, and 802.11ac system parameters is presented through a number of experiments starting from TDOA estimation using a number of cables between the transmitter and receiver to 2-D wireless indoor positioning with NLOS condition using a single transmitter and a number of BSs equipped with antenna arrays.

Chapter 7: UDP Identification for High-Resolution Wireless Indoor Positioning

The MU is often in a NLOS state, and the direct path could be completely blocked due to the harsh nature of indoor environments. Therefore, the estimated time delay of the first path should be identified either as a very weak detected direct path (DDP) or even as an UDP. Consequently, precise estimation of the channel profile parameters is not enough for high-resolution wireless indoor positioning system. However, it stays representing a key element to identify the UDP condition. Therefore, the accurate estimation of channel profile parameters and the proper modeling of DDP and UDP channel profiles have been treated and addressed to the problem of UDP identification.

Using the new parameters of 802.11ac and the previous robust algorithms improves the ability of channel profile parameters estimation of each detected path. Improving the accuracy of channel profile parameters estimation represents our key to identify and then mitigate the state of UDP. Results have been published in [59].

Chapter 8: Conclusions and Future Work

This chapter concludes the dissertation, lists out the introduced contributions, and highlights possible future work.

CHAPTER 2
Fundamentals of Position Estimation Techniques

2.1 Wireless Positioning System Architecture

The main task of the wireless indoor positioning system is to estimate the MU coordinates in a reference map using the characteristics of radio signals transmitted between the MU and a number of BSs. The characteristics of radio signals that can be measured by a positioning system are the RSS, the DOA, the TOA, and the TDOA. The nature and the expected performance of using these measurements will be discussed in this chapter. By using these techniques, the distances between the MU and every BS, or the direction of the MU with respect to each BS can be estimated. Each BS then reports its observation to the MU positioning algorithm, where the MU position in the reference map can be estimated. The performance can also be improved, because nowadays the map of the building is normally available in an electronic format [5]. Hence, the large error estimates, which lead to walls crossing or jumping through the floors, can be easily identified and eliminated based on the electronic map of the building.

The wireless indoor positioning systems have different features compared with the traditional positioning systems including harsh indoor environments, special system requirements, and probably different applications with different performance requirements. The indoor radio propagation channel characteristics are totally different from that in the traditional positioning systems such as GPS or radar. In addition, the wireless indoor channel should be modeled in a different way compared with the wireless indoor channel models that are used for telecommunications [18]. The quality of received radio signals can be improved using the diversity techniques including temporal, spatial, or spectral diversity techniques. In this work, the diversity techniques and how to implement them based on some key elements of the existing wireless infrastructure will be investigated. Increasing the number of BSs, which report the position measurements, improves the performance of positioning.

The general block diagram of a typical wireless positioning system is shown in Fig. 2.1. From Fig. 2.1, it can be observed that the performance of MU position estimation is a function in the quality of the received radio signals, type of observation and the performance of its estimation algorithm, the performance of MU coordinates estimation algorithm, the ability of UDP condition identification, and the ability of final estimation correction using the electronic floor plan. The estimated-position improvement using the floor plan will be outside the scope of this work.

In this chapter, the wireless indoor channel model will be presented. The principle of the positioning observations with their advantages and disadvantages will be presented as well as the principle of MU coordinates estimation. More details and the proposed algorithms will be presented in Chapter 5. The summary of the fundamentals of position

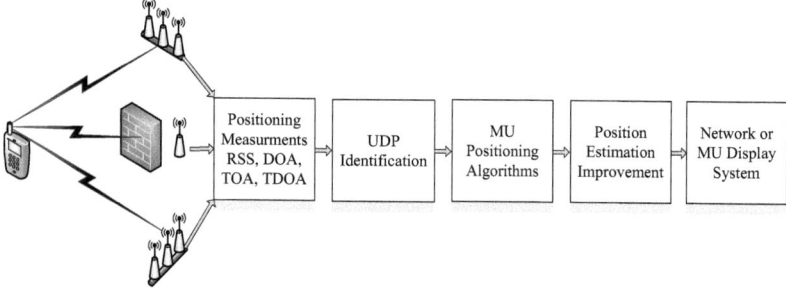

Figure 2.1: The general block diagram of a wireless positioning system based on the radio signal characteristics.

estimation techniques presented in this chapter has been introduced in [52]. By the end of this chapter, we will understand the fundamentals of wireless indoor positioning, which form a basis of this work presented in the following chapters.

2.2 Indoor Radio Propagation Channel Model

As an initial step for wireless indoor positioning system, let us start by presenting the propagation time delays and the relative DOAs of multipath propagation signals using a single transmitter and a single antenna array represented by the ULA as shown in Fig. 2.2, where the transmitter is located in the far field. To model a radio channel, only a finite number of paths are considered to approximate the real environment as illustrated in Fig. 1.2. In the subsequent analysis, we will assume a multipath channel impulse response (CIR) $h(t)$, which is given by [27]

$$h(t) = \sum_{l=1}^{L} \alpha_l a(\theta_l) \delta(t - \tau_l) \qquad (2.1)$$

where L is the number of distinct propagation paths, $\alpha_l = |\alpha_l| e^{j\phi_l}$ and τ_l represent the complex gain and the propagation time delay of the lth path, respectively. $a(\theta_l)$ is the array response vector to the lth path from direction θ_l. The array response of the mth antenna to the lth path can be represented as:

$$a_m(\theta_l) = e^{-j2\pi f \tau_m(\theta_l)} \qquad (2.2)$$

where $\tau_m(\theta_l) = m\rho \sin \theta_l / c$ represents the different propagation time that the plane wave impinging from direction θ_l needs to span the different distance between the antenna m and the reference antenna in the antenna array as shown in Fig. 2.2. The parameter ρ is the distance between adjacent antenna elements, which is equal to the half wavelength, and c is the speed of light, since the radio waves travel at the speed of light in free space or air [18]. From (2.1), the channel profile of a multipath wireless channel can be represented

2.2. Indoor Radio Propagation Channel Model

by a tapped delay module as shown in Fig. 2.3, where $x(t)$ and $r(t)$ are the transmitted and the received signals. In a mobile communication, arrival angles and time delays are relatively stationary, where the amplitude and the relative phase of each path are highly non-stationary and subject to Rayleigh fading [27]. From (2.1), the CFR representation is given by

$$H(j\omega) = \sum_{l=1}^{L} \alpha_l a(\theta_l) e^{-j\omega_c \tau_l} \quad (2.3)$$

where ω_c is the phase velocity of the carrier frequency. For positioning purposes, our concern is to estimate the time delay and the relative angle of the first path, which represents a key element in this work to reduce the complexity and to mitigate the pairing problem in case two or more paths have equal DOAs or time delays, as it will be explained in the following chapters.

2.2.1 The CFR and CIR of Multipath Wireless Channel Understanding

To understand the principle of multipath wireless channel clearly, let us record a single snapshot from a real channel and then plot its CIR and CFR. The network analyzer, Agilent ENA E5071C, was used. It was used to measure the indoor CFR at a carrier frequency of 5.25 GHz and bandwidth of 320 MHz. The measurement system is shown in Fig. 2.4. Omnidirectional antennas were used, which have 3 dBi gain and 0.668 ns time delay, measured using the wideband time-domain transmission (TDT), Agilent 86100A. The description of system equipments will be provided later in Section 6.1, while it is interesting here to show the CIR and CFR of a NLOS wireless multipath channel. As shown in Fig. 2.4, the transmitter antenna at the location point in the corridor was connected to port 1 of ENA through cable 1 of length 12.3 m and time delay 46.807 ns, measured using the TDT. At the receiver side in the lab, the receiver antenna was connected to port 2 through cable 2 of length 1 m and time delay 4.79 ns. The antenna height in both Tx and Rx was 152 cm. The transmitted power of ENA was 10 dBm.

The complex CFR can be obtained by sweeping the channel at uniformly spaced frequencies. The frequency spacing was configured to be like the subcarrier spacing of OFDM WLAN systems, which is equal to Δf =312.5 kHz. The real and imaginary parts of the forward transmission coefficient S21 were measured and stored for further processing. The measurement system has harsh NLOS environment, where the direct path has been corrupted by a circular concrete column with a diameter of 46 cm as shown in Fig. 2.4. The CIR can then be obtained using the inverse fast Fourier transform (IFFT). If the number of CFR samples is N_{FFT}, the fundamental period is given by $N_{FFT} \times T$, where T is the sampling interval. The resolution time T of IFFT is equal to the inverse of the system bandwidth $T = 1/(N_{FFT} \times \Delta f)$. The transmitted signal is received through multiple number of paths as it is obvious from the CIR plot in Fig. 2.5. These paths are caused by different objects located between antenna pairs through a number of reflections, diffractions, and scattering. From Fig. 2.5, the first detected path (FDP), which has been colored by green, has a very low amplitude compared with the other peaks of the other paths. The vertical dashed line in Fig. 2.5 denotes the actual TOA, which has been colored by red. Each path has different amplitude and phase leading to constructive and destructive interference. The multipath interference allows the wireless channel to have

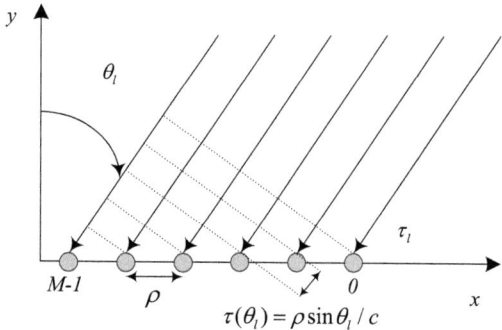

Figure 2.2: Time delay and DOA representation using a uniform linear antenna array.

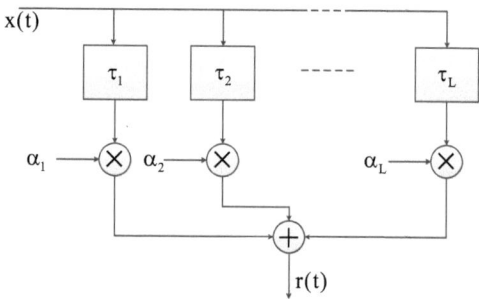

Figure 2.3: CIR representation as a tapped delay module.

2.2. Indoor Radio Propagation Channel Model

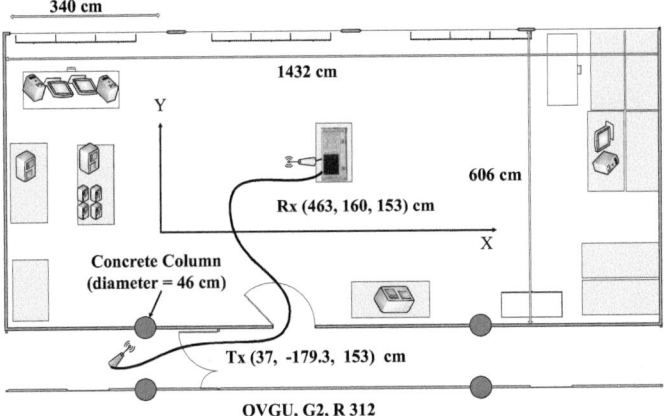

Figure 2.4: Measurement environment.

frequency-selective fading as it is clear from the recorded CFR in Fig. 2.6, where the CFR gain is not flat. It has some deep fades in the received spectrum due to destructive interference.

As a conclusion, if a high data stream is transmitted over a narrow bandwidth, and the transmission frequency is subjected to the frequency selectivity, the received signal could be lost. As a consequence, data detection is difficult due to the interference of multiple data symbols. On the other hand, using a wide bandwidth leads to a small loss in the signal power rather than a complete loss if the frequency selectivity is occurred. Examples for that are CDMA and OFDM systems.

2.2.2 Time Delay Estimation using Channel Delay Profile

In wireless positioning systems based on time delay estimation, the TOA of the direct path indicates the distance between the transmitter and the receiver. However, the TOA of the FDP of the channel profile, colored by green in Fig. 2.5, is used as an estimated time delay of the direct path denoted by $\hat{\tau}_1$. The estimated distance between antenna pair of BS and MU is equal to the time of flight (the propagation time of the transmitted signal) multiplied by the speed of propagation (the speed of light c) as

$$\hat{d} = c \times \hat{\tau}_1. \qquad (2.4)$$

Based on the resulting CIR in Fig. 2.5, the estimated time delay of the FDP is $\hat{\tau}_1 = 18.87 ns$ after calibration (removing the time delay of cables and antennas). The estimated distance between antenna pair is then $\hat{d} = 5.66 m$. If the XYZ coordinates of Tx antenna is (x_1, y_1, z_1) and of Rx antenna is (x_2, y_2, z_2), the distance between them is then calculated:

$$d = \sqrt{(x_1 - x_2)^2 + (y_1 - y_2)^2 + (z_1 - z_2)^2}. \qquad (2.5)$$

18   **Chapter 2. Fundamentals of Position Estimation Techniques**

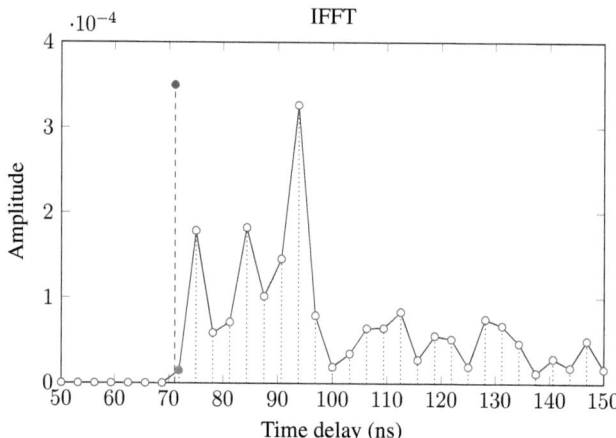

Figure 2.5: The CIR obtained using IFFT with 320 MHz BW, where the vertical dashed line represents the actual time delay, and the smallest peak colored by green is the first detected path.

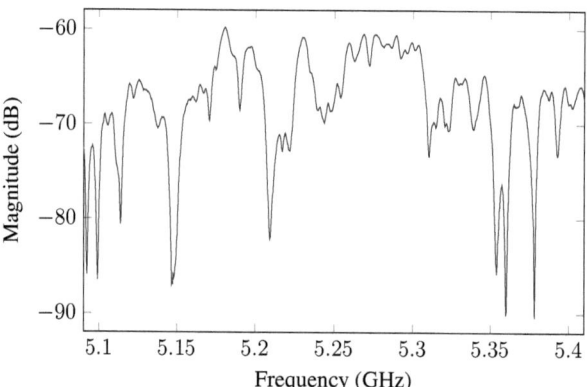

Figure 2.6: The CFR of recorded channel comprising frequency selectivity.

In our demo in Fig. 2.4, it is $d = 5.44m$. The estimation error is calculated:

$$\xi_d = |\hat{d} - d|. \tag{2.6}$$

This estimation error is known as the distance measurement error (DME). The DME of the previous experiment is $\xi_d = 22cm$, where 320 MHz bandwidth is used. If the system bandwidth is 160 MHz, the CIR obtained using IFFT is shown in Fig. 2.7. The DME ξ_d is 117 cm. However, most of multipaths in the CIR occur close together, which means the accuracy of using IFFT is poor, and it is limited to the sampling interval. Although IFFT method with padding zeros can improve sampling range resolution, the cost and complexity of calculation are too high [60]. Our goal is to be able to expand the region of interest, which represents the first a few tens of nanoseconds rather than sweeping around the whole unit circle.

An alternative method to the IFFT method with padding zeros is to use the inverse chirp Z transform (ICZT) [61]. It allows to expand the region of interest by specifying the starting time, the time resolution, and the number of time steps. If the time resolution increased by 2, the resulting CIR is presented in Fig. 2.8. The DME has been improved to 23.26 cm. However, if the time resolution increased by 4, the same DME is obtained without any additional improvement. Based on the previous experiment, we can say that the estimated distance based on the time delay measurement is usually larger than the real distance between the transmitter and the receiver.

While the time delay estimation of the FDP is the most important for wireless positioning, the average time delay of the channel profile is the most important for telecommunications. It is described by the root mean square (RMS) delay spread, which is defined as the second central moment of the channel power delay profile. It is defined as [62], [63]

$$\tau_{RMS}^2 = \frac{\sum_{l=1}^{L} (\hat{\tau}_l - \tau_{MED})^2 |\alpha_l|^2}{\sum_{l=1}^{L} |\alpha_l|^2} \tag{2.7}$$

where τ_{MED} is the mean excess delay defined as

$$\tau_{MED} = \frac{\sum_{l=1}^{L} \hat{\tau}_l |\alpha_l|^2}{\sum_{l=1}^{L} |\alpha_l|^2} \tag{2.8}$$

where $|\alpha_l|^2$ is the power of lth path, and L is the number of effective paths.

2.3 Transmitter Position Categories

In telecommunication applications, it is interested in modeling the behavior of multipath channel in terms of propagation path loss, shadow fading, Doppler spread, time dispersion, delay spread and the other parameters. On the other hand in time-based wireless indoor positioning, it is interested in studying the behavior of the direct path between the transmitter and receiver antennas, where the phase of the direct path is a function of distance between antenna pairs [15]. The MU could be in a LOS or in a NLOS with respect to the fixed BSs. In case of a LOS, the direct path is available and can be detected easily. But, in case of a NLOS, the direct path could be detected or undetected. The channel profiles can be categorized based on the availability of the direct path as [15], [14]:

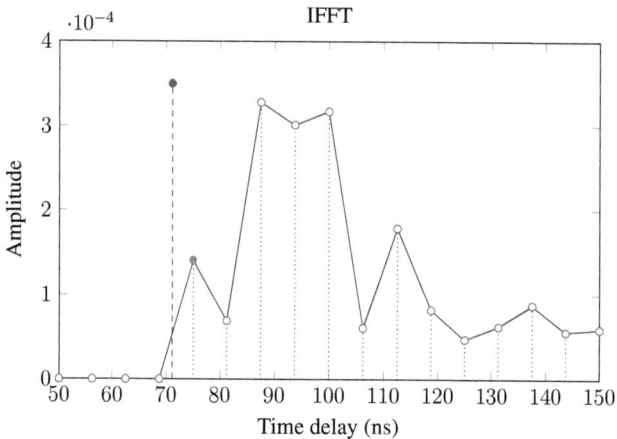

Figure 2.7: The CIR obtained using IFFT with 160 MHz BW, where the vertical dashed line represents the actual time delay, and the peak colored by green is the FDP.

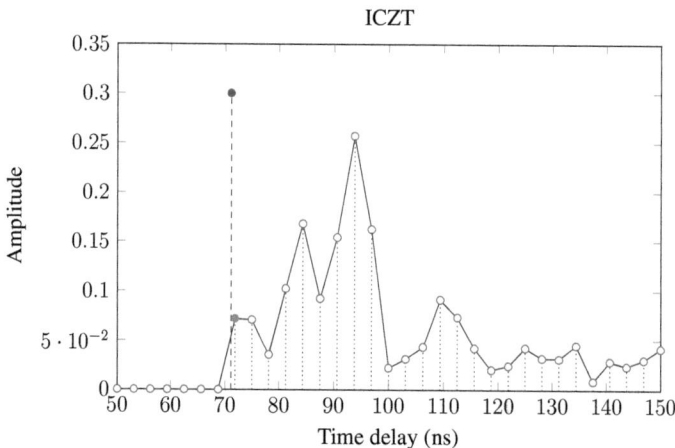

Figure 2.8: The CIR obtained using ICZT with 160 MHz BW, where the vertical dashed line represents the actual time delay, and the peak colored by green is the FDP.

1. *Dominant Direct Path*: The direct path is the strongest path in the channel profile.

2. *Non-Dominant Direct Path*: The direct path is not detected as the strongest path by the receiver, but it can be detected by an advance signal processing.

3. *Undetected Direct Path*: The direct path cannot be detected while the other paths can be detected. The FDP is assumed to be a direct path by the receiver which causes a huge DME.

4. *No Coverage*: The MU could be in a position where is no coverage, or the necessary number of BSs for positioning, which can receive a power more than the detection level, is not enough.

The UDP multipath condition can also be categorized into two types [15], [64]:

1. *Shadowed UDP (SUDP)*: SUDP multipath condition occurs when the direct path between the transmitter and receiver is blocked by a large object such as an elevator or a metallic chamber.

2. *Natural UDP (NUDP)*: NUDP multipath condition occurs in some environments when the received power of the direct path is very low and cannot be detected due to the large distance between the transmitter and receiver, but there are still other paths arriving with powers that can be detected.

The summary of the above MU position categories is shown in Fig. 2.9 [14]. In the practice, the MU state changes between those states during the MU movement. In general, the MU position can be estimated if the direct path is still detected. But if the direct path is shadowed, then none of the traditional techniques are still effective for precise indoor positioning, especially those which are based on the time delay or DOA estimation [15]. It is worth mentioning that if the MU is in a UDP state, the estimation of MU position has a very large DME. However, it can be mitigated by the assistance of the previous position of MU, and the direction of MU movement with the electronic map of the floor plan.

2.4 Bandwidth Consideration

System bandwidth is an important parameter that affects on the performance of the time-based wireless indoor positioning system. As it has been explained, the time-based systems measure the distance between a transmitter and a receiver based on the estimated propagation time delay. The TOA can be measured by either measuring the phase of a received narrowband carrier signal or directly by measuring the arrival time of a wide-band narrow pulse [18]. The wireless positioning techniques based on the time delay estimation can be grouped based on the system bandwidth into three categories as [18]: narrowband signal positioning techniques, wideband signal positioning techniques also known as super-resolution techniques, and UWB signal positioning techniques.

In the narrowband positioning techniques, the distance is measured by the phase difference between received and transmitted carrier signals. The TOA of the signal, τ, and the phase of the received carrier signal, ϕ, are related by, $\tau = \phi/\omega_c$. However, using a narrowband carrier signal in indoor positioning is not like the differential GPS, where the

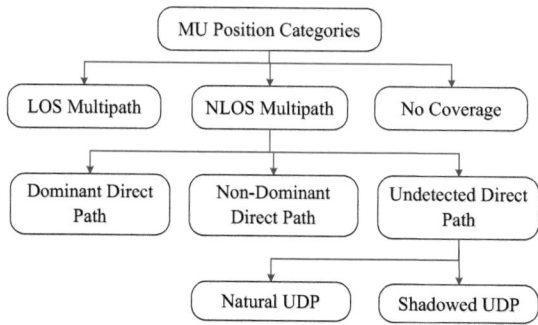

Figure 2.9: The MU position categories based on the availability of the direct path [14].

direct path of a transmitted signal is available most of the time. In indoor environments, there are a number of effective paths each has different amplitude and phase. Therefore, the distance cannot be estimated accurately using a narrowband carrier signal in a heavy multipath environment [18].

In wideband signals, a known pseudo noise (PN) sequence for example is used by a transmitter such as the direct sequence spread spectrum (DSSS) wideband signal. The same PN sequence is generated also in the receiver, which is correlated with the received signal. The arrival time of the first correlation peak is used to find the distance between a transmitter and a receiver. However, the accuracy of this method is limited due to the complexity of multipath indoor propagation channel. Usually, the first correlation peak does not represent the actual peak, and the resolution of TOA estimation is based on the base width of the PN correlation function. Therefore, a number of researchers have studied frequency-domain super-resolution algorithms, which can be used to estimate the time delay with high resolution from the CFR.

In UWB, signal bandwidths that exceed one GHz are used in the unlicensed ranges such as 2 to 3 GHz or 57 to 66 GHz. Using those large bandwidths means extremely short duration pulses. The UWB signal is not much affected by the multipath fading. Therefore, the high time resolution of the UWB signal makes positioning based on the TOA is very accurate. Recently, there are a number of proposed techniques which utilize the UWB signal for indoor positioning applications such as in [40], [65], [66].

In indoor environments, there are many objects in the surrounding area, hence, there are many paths close to the direct path. Therefore, the peak of the channel profile is shifted from the actual TOA, resulting in a TOA estimation error, and then DME [15]. By increasing the system bandwidth, the channel profile can be split to a number of effective paths, in another way, the pulses arriving from different paths become narrower, and then the estimated TOA of the FDP is closer to the actual TOA of the direct path. To provide a reasonable protection against extensive multipath environments in indoor areas, there are two techniques employed for indoor positioning as follows [15]:

1. Bandwidths on the order of several hundred megahertz should be used. The sufficient bandwidth for accurate indoor positioning based on the TOA technique is larger than 200 MHz [4] for the receiver to be able to resolve the multipath components.

2. Using super-resolution algorithms for post processing can reduce the bandwidth requirements.

Based on the previous discussion, the accuracy of channel profile parameters estimation depends strongly on the system bandwidth. However, in practice the channel bandwidth is limited. Hence, the principle of super-resolution algorithms represents the interested option in this work and it will be presented in the following chapters. It is worth mentioning that increasing system bandwidth can decrease the DME if the direct path is still detected, but if the direct path is totally blocked, DME is not necessary reduced [67].

2.5 Main Sources of Error in Time-based Wireless Indoor Positioning Systems

In time-based wireless indoor positioning, the time delay of the FDP is used as an estimated time delay of the direct path denoted by $\hat{\tau}_1$. To investigate $\hat{\tau}_1$ features, let us assume the MU and the BS are synchronized, although using TDOA observations represents our concern. The estimated distance between the MU and the BS is then given by (2.4) as, $\hat{d} = c \times \hat{\tau}_1$. The DME is then given by (2.6) as, $\xi_d = |\hat{d} - d|$, where d is the actual distance. The main sources of the DME in time-based wireless indoor positioning have been reported in [14]. The first source is the presence of rich multipath channel, which produces an extra time delay in the FDP estimate compared to the actual direct path. The second source is the blockage of the direct path by a large object such as a chamber or an elevator, or due to the large distance between the MU and the BS. The absence of the direct path leads to a very large DME. The third source is the induced propagation time delay, where the speed of the radio waves varies in different media. Therefore, the estimated time delay of the FDP including the three types of time error is

$$\hat{\tau}_1 = \tau_1 + \tau_m + \tau_{UDP} + \tau_{pd} \tag{2.9}$$

where τ_1 is the propagation time delay of the direct path, τ_m, τ_{UDP}, and τ_{pd} are the time errors due to the rich multipath channel, the blockage of the direct path, and the induced propagation time delay, respectively. In this work, the principle of subspace algorithms is used, which can provide a reasonable protection against sever multipath environments, and the parameter τ_{pd} can be assumed as an insignificant parameter. The largest DME is coming from the blockage of the direct path. Therefore, the problem of UDP condition identification should also be investigated to mitigate the estimated parameters of the UDP channel profile. The main sources of the DME investigation with UDP identification have been presented in [59], as it will be described in Chapter 7.

2.6 Radio Signal Characteristics

The three basic properties that enable distance or direction measurement for MU position estimation by analysis the received radio signals are the RSS, the DOA, the TOA, and the TDOA. The principle of these positioning techniques with their advantages and disadvantages will be presented in the remaining of this chapter.

2.6.1 Received Signal Strength (RSS) based Techniques

The power density of an electromagnetic wave is proportional to the transmitted power and inversely proportional to the square of the distance to the transmitter [6]. As a result, the RSS at the receiver is related to the distance between the transmitter and the receiver. This relation as well as the combination between waves that reach a receiver over different paths are the basis for distance estimation. The RSS can be obtained and reported from the NIC that is available in most wireless devices. For example in IEEE 802.11 standards, the MAC layer provides the RSS for all active access points in a quasi-periodic beacon signal [7]. Wireless sniffer tools provide an access to MAC address and RSS values of WLAN access points. Therefore, a positioning system can be implemented on top of existing WLAN infrastructures without the need for any additional hardware [7]. Using RSS for positioning is known as the RSS-based ranging technique. In general, RSS techniques for positioning can use the principle of path loss model [68], [23], finger printing [8], compressive sensing [69], [70], and propagation modeling [9], [10], [11]. The principle of each method will be presented in the following.

2.6.1.1 Path loss Model based Techniques

In free space, the parameters that directly affect the relationship between received power P_r and distance d at wavelength λ are included in the Friis equation [71]

$$P_r = \frac{P_t G_t G_r \lambda^2}{(4\pi d)^2} \qquad (2.10)$$

where G_t and G_r are transmitter and receiver antenna gains. If the radiated power ($P_t G_t$) of the transmitter is known at the receiver, the distance between the transmitter and the receiver can be calculated. However, in indoor environments there are many objects in the vicinity of the transmission path, which can change the relationship between received power and distance. The received signal is a combination between a number of paths, each path has its own interaction with the nearby objects in the signal path. As a result, the relationship between the RSS and the distance between the transmitter and the receiver is presented mathematically in a form called the path loss model [68]. The mean path loss increases exponentially with the distance [68], hence, the mean path loss is a function of distance to the n power. The mean path loss model derived from the log-normal shadowing model is [68]

$$\overline{PL}(d)[dB] = PL(d_0) + 10n \times \log_{10}(d/d_0) + \chi_\sigma \ [dB] \qquad (2.11)$$

where $\overline{PL}(d)$ is the mean path loss, n is the mean path loss exponent which indicates how fast path loss increases with distance, d_0 is a reference distance, d is the distance between the transmitter and receiver, and χ_σ is a zero mean log-normally distributed random variable with standard deviation σ in decibels. The $PL(d_0)$ is due to free space propagation to the reference distance, for example, $d_0 = 1m$. The mean path loss exponent n and standard deviation σ (in decibels) are viewed as parameters that are a function of building type and building layout, which are between the transmitter and receiver. This model predicts path loss as a function of distance accurately when the model parameters n and σ are determined as a function of the general surroundings precisely, which is impossible.

By measuring the RSS at the receiver, where the path loss model and the reference power are known a priori, the distance between antenna pairs can be estimated. To estimate the MU position, a number of fixed BSs should be used, a minimum of three, as shown in Fig. 2.10. From Fig. 2.10, each distance measurement determines a circle geometrically, which is centered at the reference BS. It should be noted that the radii of the solid-line circles represent the estimated distances between the MU and BSs, where the radii of the dashed-line circles represent the real distances between the MU and BSs. The estimation of distances based on the RSS measurements are not accurate, hence, the intersection of those three circles makes a region of the possible MU position as shown in Fig. 2.10. It is called the region of uncertainty [6]. Furthermore, it should be noted that the estimated distances could be smaller or larger than the actual distances as shown in Fig. 2.10. Therefore, more than three BSs are needed to improve the accuracy of positioning. The principle of MU coordinates estimation based on the path loss model will be presented in Section 2.6.3.2.

To measure the ability of a distance estimation using the path loss model, some experiments have been presented in Section 6.2.1. In indoor environments, the materials of walls and floors, number of floors, layout of floors including the size of rooms, and the position of obstructing objects have a significant effect on the path loss. As a result, it is difficult to find a model applicable to all environments [7]. In addition, the value of RSS is instantaneous and it varies over the time, even at a fixed position as it is obvious from the experiments of Section 6.2.1; this is due to shadow fading and multipath fading as well as the NLOS caused by humans, walls, and many objects inside the building [7]. Hence, an immediate conclusion is that using RSS technique based on the path loss model is not the proper option for accurate indoor positioning system. Therefore, it will be ignored in our future work. However, it can be used in indoor positioning systems, which require low accuracy.

2.6.1.2 Fingerprinting based Techniques

As an alternative method based on measuring RSS is the fingerprinting or sometimes called a pattern recognition. The basic principle of fingerprinting is to give each position in a building a unique set of signal strengths received from the properly distributed BSs. The position signature is a vector $\mathbf{R}_k$, which is the average measured RSS from I BSs at position k in the area of interest; i.e. $\mathbf{R}_k = \{r_1, r_2, .., r_I\}$, where r_i is the average measured RSS from BS number i at position k. The database of these position signatures with their coordinates is known as a radio map. The generation of the radio map is done in offline mode, which is called the initial training phase. In online mode, a comparison between the RSS observations to radio map signatures is done and return a position estimate. The position estimate can be obtained using for example the Euclidean distance,

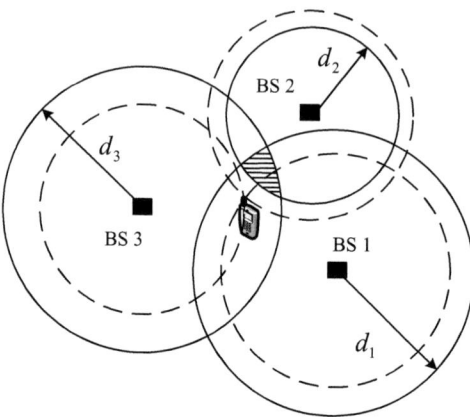

Figure 2.10: Wireless positioning system based on distance estimation using RSS technique.

which finds the most closely signature to the observation, where the estimated position is assumed that it has the coordinates of the best matched signature in the radio map. To get the principle and the required effort of fingerprinting more clearly, a simple experiment has been made in Section 6.2.2.

From the mentioned experiment, it can easily find the following challenges to create the radio map. The number of points with their locations, which are strongly depended on the required accuracy, and the number of time samples per each point should be determined. The expected accuracy depends on the resolution of radio map creation process; the minimum distance between points of radio map as shown clearly in Section 6.2.2. In addition, if the area of interest is large, the necessary number of BSs increases, and then the required effort of radio map creation and the size of the database increase dramatically. Therefore, searching for the best match inside this large database requires a huge computational burden. Furthermore, it is difficult sometimes to distinguish between the positions with the same signature [5].

To reduce the complexity of positioning based on fingerprinting, some preprocessing can be made before making the actual processing. The radio map can be clustered to some zones to reduce the computational burden and to improve the performance of the positioning system as proposed in [72]. Some efforts have also been made in the literature to reduce the effects of RSS variations due to channel impediments by using a compressive sensing (CS) principle [73]. It is called also the compressive sampling, which offers accurate recovery of sparse signals from a small number of measurements. The principle of using CS is that position estimation is a sparse problem. According to the CS theory, the accuracy of position estimation can be improved using only a small number of noisy measurements as presented in [69], [70].

The major drawback of fingerprinting techniques is that for any major change in the indoor environment, measurements of the radio map should be repeated. Furthermore, the RSS observations in the offline mode could be different from those in the online mode.

2.6. Radio Signal Characteristics

For example, the user body spreads the range of RSS values by a significant amount in the range of 3 dBm, and the orientation of user leads to the attenuation by amount of 9.3 dB due to the obstruction from the body as reported in [8].

It can be noted from the previous that the fingerprinting method is a tedious process. Therefore, some methods have been suggested in the literature to reduce the effort and cost of the indoor positioning system such as RT (ray tracing) for predicting the CIR. Ray tracing represents the electromagnetic waves as rays and produces deterministic channel models that operate by processing user-defined environments. Due to the site-specific nature of indoor environments, RT process is implemented by considering the position, the orientation, and the electrical properties (including: permittivity, conductivity and thickness) of individual walls and objects that are possible in a given area. More details on the RT principle can be found in these references [9], [10], [11], [74].

The summary of improvements to use RSS principle for indoor positioning is presented in Fig. 2.11. The conclusion of the above section is that using RSS based ranging techniques are not the proper option for accurate indoor positioning system. However, the RSS can be applied for indoor positioning systems that require low accuracy. Therefore, using RSS techniques will be ignored in our future work.

2.6.2 Direction of Arrival (DOA) based Techniques

The DOA or the angle of arrival (AOA) technique is probably the oldest technique for wireless positioning [6]. It is based on using the directional property of directional antennas and recently using antenna arrays. The direction of a radio wave can be estimated by observing the RSS changes while varying a known spatial radiation pattern of the receiving or transmitting antenna. The DOA may be determined to be the point in the pattern rotation where the RSS is maximum, or where a null in RSS occurs, depending on the reference point [6]. However, for antenna arrays, there is a relationship between the direction of a signal and the associated received steering vector as shown in Fig. 2.2 and equation (2.2). There is a difference in the received phase at each element in the antenna array, which comes from the different distance between each antenna element and the reference antenna. Therefore, it is possible to estimate the direction of a signal from the received signals at antenna elements.

The MIMO technology in recent standards of 802.11 could be used for DOA estimation. The antenna array could be ULA, URA, or circular array. In this work, only the

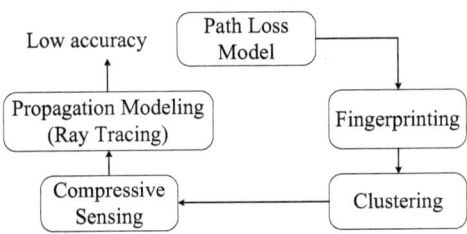

Figure 2.11: The summary of RSS measurement techniques.

ULA is used as shown in Fig. 2.2. The antenna array is typically installed at the BS, because it is difficult to employ it inside the MU. Therefore, the DOA technique is normally employed in wireless indoor positioning systems that are based on the multilateral system (network-based architecture), which represents the interesting option in this work. The MU device should be very simple. Using antenna arrays will be employed in this work for two aspects: DOA estimation and spatial diversity to improve the quality of received radio signals.

The DOA technique uses simple triangulation to estimate the position of MU, where two BSs are enough to do that as shown in Fig. 2.12. The DOAs at two BSs are θ_1 and θ_2. The coordinates of MU (x, y) can be found easily from the trigonometry as

$$x = \frac{x_2 \cdot \tan(\theta_2) - y_2}{\tan(\theta_2) - \tan(\theta_1)}, \quad y = x \cdot \tan(\theta_1). \tag{2.12}$$

In practice, the DOA cannot be measured exactly as shown in Fig. 2.12. There are some uncertainties in the measurements of θ_1 and θ_2 as, $\pm \Delta \theta_1$ and $\pm \Delta \theta_2$, respectively. The estimated position of MU are then located in the overlapping region of the two beams. The accuracy of DOA estimation depends strongly on the size of the region of uncertainty. Hence, if the number of antenna elements increases and a robust algorithm has been used, the accuracy of MU coordinates estimation increases. From Fig. 2.12, we can also observe that the accuracy of MU position estimation degrades if the distance between the MU and BS increases. In addition, the position of MU relative to fixed BSs is very important in determining the accuracy of estimated position. The highest accuracy could be achieved if the positions of MU and BSs form an acute triangle (all angles are less than 90°) [6]. Therefore, more BSs should be used to improve the accuracy of estimation.

There are many techniques in the literature for DOA estimation of narrowband signals using antenna arrays including maximum likelihood (ML) estimator [75], and super-resolution subspace techniques such as Root MUSIC [30], ESPRIT [28], and MP [34]. The DOA technique can be combined with RSS such as in [76] or with TOA such as in [31] for narrowband signals and in [40] for IR-UWB systems, or with TDOA such as in [33] for cellular systems. The DOA can also be estimated by measuring the phase difference between antenna elements in case of using UWB systems such as in [65], [66]. More details and comparisons to this work will be presented in the following chapters.

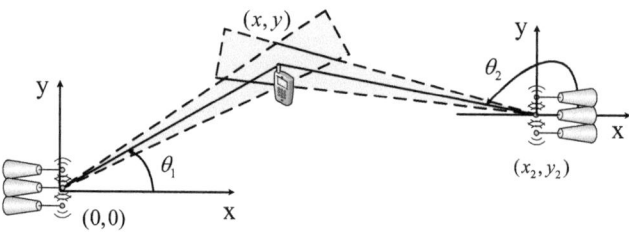

Figure 2.12: Triangulation illustration using DOA technique in 2-D scenario.

2.6. Radio Signal Characteristics

Finally, the advantages of using DOA are summarized in the following:

1. It does not require synchronization between BSs such as in time-based techniques. The knowledge of the transmitted power and the reference power are not required such as in RSS.

2. The loci are lines. Hence, the MU position can be calculated easily by triangulation.

3. Antenna array should be installed in the BS, which is provided by modern wireless systems such as IEEE 802.16, IEEE 802.11n, and IEEE 802.11ac, where the MIMO technology is employed.

4. Knowledge of DOA can provide other features for telecommunications besides positioning such as interference cancellation, and space-division multiplexing [77].

The disadvantages of using DOA are summarized in the following:

1. It increases the complexity of the positioning system, where an antenna array should be installed at each BS.

2. The accuracy of MU position estimation degrades if the distance between the MU and BS increases.

2.6.3 Signal Propagation Time Delay based Techniques

The time-based wireless positioning systems use the propagation time of the signals as an estimate for the distance between the transmitter and receiver antennas as shown in Section 2.2.2. The time-based techniques are not like RSS-based techniques, which are affected by large distances, orientation of the wireless device, and the value of RSS, but they are affected by system bandwidth, multipath effects, and SNR. The principle of time-based techniques is somehow like in case of GPS, but the received signals from satellites are affected only by SNR. However, in wireless indoor positioning, the wireless indoor environments have harsh and challenging propagation situations.

The time-based wireless positioning techniques can be categorized into the following:

1. *TOA techniques*: In which the MU position can be determined from the TOA observations at a number of fixed positions, where the transmission time is known. Hence, clock synchronization is required on both sides of the communication link, the MU should be synchronized to the network by somehow method, which represents the major disadvantage of the TOA techniques.

2. *TDOA techniques*: In which the MU position can be determined from the differences of reception time at a number of fixed positions. Hence, clock synchronization is required only on one side of the communication link, the side of the fixed BSs.

3. *Phase of Arrival (POA) techniques*: In which the phase of the received signal is related to time and distance through the signal wavelength and speed of light as, $\phi = \omega_c d/c$ [18].

2.6.3.1 Indoor Positioning Techniques based on TOA

A variety of signal processing techniques are available for time delay estimation in the literature. In [4], a frequency-domain TOA estimation has been proposed using MUSIC algorithm. It is used to transform the measured CFR to time-domain pseudospectrum. The TOA can be obtained by detecting the first peak of the pseudospectrum in the delay axis using a peak detection algorithm. Similarly in [20], the TOA has been estimated using Root-MUSIC algorithm based on the measured CFR using IEEE 802.11a/g system parameters. A comparison between MUSIC and ESPRIT for time delay estimation using the estimated CFR has been investigated in [25] using IEEE 802.11b standard parameters; the ESPRIT algorithms can achieve a better accuracy level than MUSIC algorithms. In [78], the position of first arrival path in wireless OFDM systems is estimated from the estimated CIR. The characteristics of the information theoretic criteria are exploited to estimate the TOA of the first path. However, the accuracy of the first arrival path estimation depends on the sampling interval.

The TOA can be used with the other techniques. For example in [31], the joint estimation of time delays and DOAs of narrowband signals has been proposed using the estimated CIR, where the transmitted pulse shape function is known. The 2-D ESPRIT is used to separate and estimate the phase shifts due to the delay and direction of incidence. Similarly in [66], the joint estimation of TOA and DOA for UWB systems has been proposed. The estimation of TOA is performed in the frequency-domain based on the power delay spectrum computation by means of a fast Fourier transform (FFT) calculation. The DOA estimation is obtained from independent TOA measurements at each antenna element by means of a linear estimator.

Other studies proposed algorithms to estimate the TOA based on the time delay of successful transmission such as in [79]. Similarly in [80], the joint TOA and RSS is proposed for WLAN positioning system, where the TOA has been estimated from the time delay of successful transmission, and the path loss model has been used to calculate the distance between the MU and the BS. Although the time delay estimation based on successful transmission does not require any hardware modification in the existing network, the accuracy of positioning based on this principle is very low in the range of 3.96 meters. In order to get an accurate TOA-based indoor positioning system, the occurrence of direct path blockage should be identified and mitigated. To identify the UDP condition, some algorithms have been presented in [15], [81], [82]; more details to the problem of UDP identification will be given in Chapter 7.

2.6.3.2 2-D Wireless Positioning based on TOA

The principle of MU coordinates estimation based on the distance estimation using TOA is like that of using the path loss model as shown in Fig. 2.10. In both cases, the distances d_1, d_2, and d_3 are estimated by measuring the signal propagation time delay or the RSS at each BS. However, the estimated time delay is usually larger than the actual time delay, but the measured RSS variates around the actual RSS; it could be smaller or larger than the actual RSS. Geometrically, each estimated distance determines a circle, centered at the BS.

Fig. 2.13 shows the geometric of 2-D wireless positioning system with one MU and three BSs, all of them are located in the same plane. The equations of the three circles in

2.6. Radio Signal Characteristics

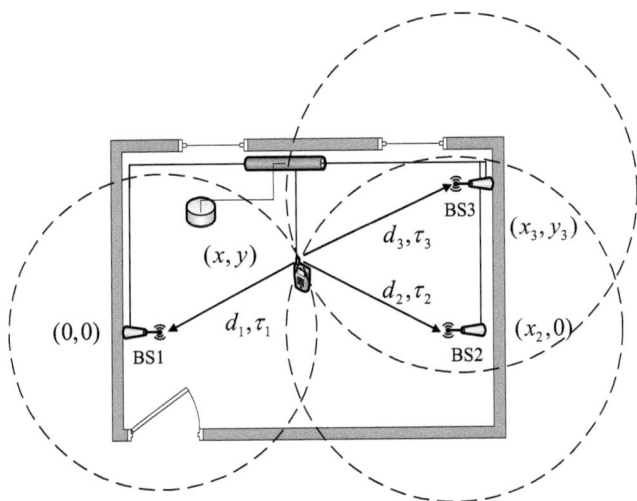

Figure 2.13: 2-D TOA positioning system using three receivers.

Fig. 2.13 are
$$d_1{}^2 = x^2 + y^2$$
$$d_2{}^2 = (x - x_2)^2 + y^2 \qquad (2.13)$$
$$d_3{}^2 = (x - x_3)^2 + (y - y_3)^2$$

The solution of those three nonlinear equations represents the estimated MU coordinates as

$$x = \frac{x_2^2 + d_1^2 - d_2^2}{2x_2}, \qquad (2.14)$$

$$y = \frac{x_3^2 + y_3^2 + d_1^2 - d_3^2 - 2 \cdot x \cdot x_3}{2y_3}. \qquad (2.15)$$

From Fig. 2.13, it is clear that the minimum number of BSs to estimate the MU position without ambiguity is at least three fixed BSs. By using the intersection of the three circles, the estimated position is unique and has no ambiguity. However, if the estimated distance is smaller than the real distance, three BSs (three circles) may not be able to get the region of uncertainty, which includes the MU position [5]. Therefore, more than three BSs should be used to improve the positioning accuracy. From the system setup view, the accuracy of positioning depends on the synchronization between system elements, the geometry of BSs, and the number of BSs in the positioning area.

A 2-D layout of four BSs with known coordinates is shown in Fig. 2.14. The circles have been formed where the fixed BSs are at their centers, and the radii equal to the estimated distances between the BSs and the MU, $\hat{d}_i = c \times \hat{\tau}_i$. If we assume that the true distances $\{d_i; i = 1, \ldots, 4\}$ are estimated perfectly, the MU coordinates are then the point of intersection of the circles as shown in Fig. 2.13. However, in practice, the circles do not cross at one point, because the estimated distances $\{\hat{d}_i; i = 1, \ldots, 4\}$ are not exact.

Therefore, it is necessary to define a criterion to estimate the MU position coordinates. The equations of the four circles defined by the coordinates of the fixed BSs are

$$\hat{d}_i^2 = (x_i - \hat{x})^2 + (y_i - \hat{y})^2 \;; i = 1, \ldots, 4. \tag{2.16}$$

The position of MU can be estimated using the LS error criterion. The estimated coordinates of MU $(\hat{x}, \hat{y})$ should minimize the following function [6]

$$F = \sum_{i=1}^{4} (\sqrt{(x_i - \hat{x})^2 + (y_i - \hat{y})^2} - \hat{d}_i)^2. \tag{2.17}$$

There are many algorithms to solve the nonlinear expression of (2.17), some of them are time consuming and inconvenient for implementation. However, a closed form solution to the estimation problem can be used as in [6]. The first step is to expand the factors on the right side of (2.16), and then subtract the equations of $\hat{d}_2$ to $\hat{d}_4$ from that of $\hat{d}_1$ to get

$$(x_1 - x_i)\hat{x} + (y_1 - y_i)\hat{y} = \frac{1}{2}(x_1^2 - x_i^2 + y_1^2 - y_i^2 + \hat{d}_i^2 - \hat{d}_1^2) \tag{2.18}$$

where $i = 2, \ldots, 4$. The above set of equations is an overdetermined set of linear equations in $(\hat{x}, \hat{y})$. It can be expressed in matrix form as

$$\mathbf{A}.\mathbf{z} = \mathbf{b} \tag{2.19}$$

where

$$\mathbf{A} = \begin{bmatrix} x_1 - x_2 & y_1 - y_2 \\ x_1 - x_3 & y_1 - y_3 \\ x_1 - x_4 & y_1 - y_4 \end{bmatrix}, \tag{2.20}$$

$$\mathbf{b} = \frac{1}{2} \begin{bmatrix} x_1^2 - x_2^2 + y_1^2 - y_2^2 + \hat{d}_2^2 - \hat{d}_1^2 \\ x_1^2 - x_3^2 + y_1^2 - y_3^2 + \hat{d}_3^2 - \hat{d}_1^2 \\ x_1^2 - x_4^2 + y_1^2 - y_4^2 + \hat{d}_4^2 - \hat{d}_1^2 \end{bmatrix}, \tag{2.21}$$

and

$$\mathbf{z} = \begin{bmatrix} \hat{x} \\ \hat{y} \end{bmatrix}. \tag{2.22}$$

The closed form LS solution to (2.19) is [83], [84]

$$\mathbf{z} = \begin{bmatrix} \hat{x} \\ \hat{y} \end{bmatrix} = (\mathbf{A}^T \cdot \mathbf{A})^{-1} \cdot \mathbf{A}^T \cdot \mathbf{b}. \tag{2.23}$$

The previous algorithm can be extended to a larger number of BSs. It can also be extended to 3-D problem if the number of BSs is enough.

A major drawback of the wireless positioning based on TOA is that all system elements should be synchronized with an acceptable accuracy in the range for example of one nanosecond in order to achieve a positioning accuracy less than one meter. Such accuracy is impossible to achieve with the commercial of the shelf (COTS) WLAN devices [85], [86], because it requires highly stable oscillators and robust hardware for time stamping, which is not acceptable to include inside the MU [13]. Furthermore, the hardware and software of both BSs and MU should be modified, where our goal is to design an accurate positioning system based on the existing WLAN networks with lowest modification probably in the side of fixed BSs. The MU should be a very simple wireless device.

2.6. Radio Signal Characteristics

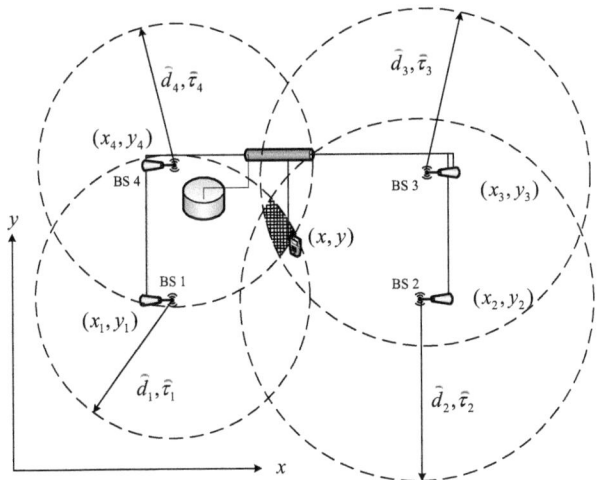

Figure 2.14: 2-D TOA positioning system using four receivers.

2.6.3.3 Indoor Positioning Techniques based on TDOA

Instead of measuring the time delay of a transmission between the MU and the BS as in TOA techniques, the difference in the time delays between the MU and a pair of fixed BSs is measured in TDOA techniques. Clock synchronization is required only on the side of the fixed BSs. Therefore, at least one additional fixed BS is required for TDOA per dimension compared with the number in case of TOA, in another way, the number of TDOA measurements should be equal to or more than the number of unknowns (order of system coordinates). Hence, for a 2-D positioning scenario, a TDOA system needs at least three fixed BSs and to estimate 3-D coordinates of the MU, it needs at least four fixed BSs [43].

Such as in TOA-based positioning system, the propagation time delay differences can be estimated based on cross-correlation, time stamps, or super-resolution algorithms. Using cross-correlation to estimate TDOA is the simplest method. The transmitted signal from the MU is received simultaneously at a number of BSs. Then, a sliding correlator is used to do cross-correlations between the received signals. The time delays between BSs can be obtained from the instants of output peaks of a sliding correlator. The major disadvantages of this method are the dependency between the resolution of the cross-correlation peaks and the sampling rate, sensitive to low SNR and multipath effects, and leading to a significant network load. In addition, finding the correlation peak in presence of carrier frequency offset results a computational task [13].

Wireless indoor positioning using WLAN can be done also using accurate time stamps such as in [13], [87]. However, the time stamp itself does not represent the arrival time of the known preamble at the antenna, but it represents the time of the preamble or any reference symbol detection. It means that the detection time includes the analog signal processing time delay due to analog amplification, mixers, and filters, and the digital signal processing time delay due to the process between the output of analog to digital converter and time of detection. The calibration of those time delays especially that coming from the analog signal processing is a challenge of using time stamps for wireless positioning. In addition, using the time stamp is highly affected by multipath fading. The mean peak power of the composite signal (the direct and the reflected signals) could be shifted; consequently, the detection time or the time stamp is shifted according to that. An investigation of the time stamp accuracy of a COTS WLAN chipset can be found in [85], [86]. The clock synchronization was performed using the IEEE 1588 protocol over IEEE 802.11b. The conclusion was that the physical layer of the wireless network can be used for a synchronization system with an accuracy of several hundred nanoseconds [86]. Obviously, it can be used in many control and instrumentation applications, but it is not designed to perform time stamping for wireless positioning.

In fact, using TDOA principle for wireless positioning can operate with system elements using their normal communication protocol, which gives it more applications than TOA. For TDOA measurements, a special message for the purpose of positioning such as time stamp is not necessary to be known, for example by including it in the transmission frame. However, a transmission frame should have a training sequence that can be recognized by the receivers. As a result, we preferred to use TDOA techniques rather than TOA techniques for wireless indoor positioning using IEEE 802.11 standards, where the preamble of the OFDM frames will be used for that, as it will be explained in Chapter 3.

As it has been described in Section 2.6.3.1, the principle of super resolution algorithms, which represents the preferred option in this work, is used to estimate the time delays of multipath channel. The robust super-resolution algorithm for TDOA and DOA estimation will be identified and explained in the following chapters. Similar to the DOA estimation problem, the multilateral system (network-based architecture) is preferred to use in this work for TDOA estimation problem, where the MU should be a very simple device.

The geometric model for estimating the position coordinates of MU using TDOA is the intersection of hyperbolas in 2-D and the intersection of hyperboloid in 3-D [6], [43]. The hyperbola is characterized by the fact that the difference in distance, $d_{12} = d_1 - d_2$, between any point on it and the two foci is constant as shown in Fig. 2.15. The difference value is positive if the point is located on the right branch of the hyperbola and negative if it is located on the left branch. If a BS is located at one focus of the hyperbola and another BS at the other focus, then the MU position is on the hyperbola [6], [88]. Based on Fig. 2.15, the distances are expressed as follows

$$d_1 = \sqrt{y^2 + (x + \frac{D}{2})^2} \qquad (2.24)$$

$$d_2 = \sqrt{y^2 + (\frac{D}{2} - x)^2} \qquad (2.25)$$

where (x,y) are the coordinates of the MU, and D is the distance between the two BSs, which are located on the x axis and they have the same distance from the origin of the x axis. The equation of the first hyperbola that defines the locus of the MU is then

$$d_{12} = d_1 - d_2 = \sqrt{y^2 + (x + \frac{D}{2})^2} - \sqrt{y^2 + (\frac{D}{2} - x)^2} \qquad (2.26)$$

2.6. Radio Signal Characteristics

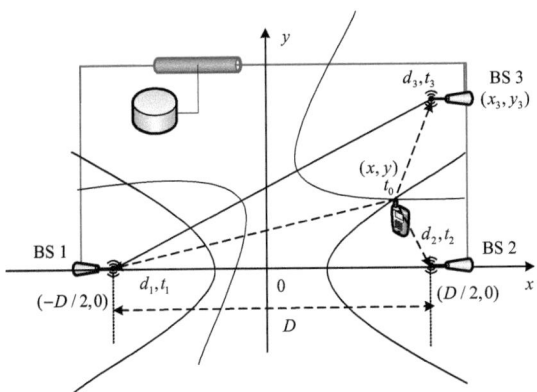

Figure 2.15: The geometric relationship between a MU and three fixed BSs in a TDOA system.

where d_{12} represents the TDOA times the speed of light as, $d_{12} = c \times \Delta t_{12}$. The general equation of the hyperbola is

$$\frac{x^2}{a^2} - \frac{y^2}{b^2} = 1 \qquad (2.27)$$

where a and b could be expressed in terms of the known quantities as, $a^2 = (d_{12}/2)^2$, and $b^2 = (D/2)^2 - a^2$.

The intersection of two or more hyperbolas that are defined from TDOA measurements represents the MU position. Although the number of independent TDOA values obtainable from I BSs is $I - 1$, we can use in a noisy environment additional pairs of measurements that are not independent, since the noise that is not correlated between those pairs gives them a degree of independence [89]. The total number of TDOA observations, Q, obtainable from I BSs is [6]

$$Q = \frac{I!}{2(I-2)!} = \frac{I(I-1)}{2} \qquad (2.28)$$

where ! is the factorial operator.

Finally, the advantages of using TDOA are summarized in the following:

1. It is preferred over using RSS and DOA, when high-accuracy positioning is required.

2. Clock synchronization is only required among the BSs, which are usually connected to a wired backbone. As a result, the synchronization between the BSs is easy to set up.

3. The time-based techniques (TOA or TDOA) are not like RSS-based techniques, which are affected by large distances, orientation of the wireless device, and the value of RSS.

The disadvantages of using TDOA are summarized in the following

1. Accurate clock synchronization between BSs is required.

2. If the direct path is totally blocked (the UDP condition), the exact time measurement will be not possible, as it has been explained in Section 2.3, where the FDP is assumed to be a direct path by the receiver which causes a large DME. Therefore, the problem of UDP identification will be investigated in Chapter 7 to mitigate its effects.

3. The time-based techniques (TOA or TDOA) are sensitive to the system bandwidth. Therefore, super-resolution algorithms will be used in this work to compensate the bandwidth limitations of IEEE 802.11 standards.

From the above, using TDOA techniques requires synchronization between BSs, which can be achieved by using a hardware support. However, this is the only way to achieve a wireless positioning system with high accuracy in harsh environments.

CHAPTER 3

System Model and Time Delay Estimation using 1-D Matrix Pencil Algorithms

In this chapter, we present some techniques for estimating the TDOA associated with signals in a multipath communication channel for wireless indoor positioning. Recently, the UMP and the BMP algorithms have been presented to estimate the DOA of coherent or non-coherent signals using the ULA. We have applied these algorithms in a new way to estimate the propagation time delay from the estimated CFR using OFDM systems. The complexity of various MP algorithms is investigated using the 802.11ac standard, and compared to the corresponding complexity of using 802.11n and 802.11a. The principle of using wideband orthogonal multi-carrier signals and diversity techniques such as the spatial diversity, spectral diversity, and the temporal diversity are presented.

The 1-D MP algorithm was presented in [34] for estimating frequencies and damping factors of exponentially damped and / or undamped sinusoids in noise using the ULA. The 1-D UMP algorithm was developed in [90] and [91] to estimate the DOA of the narrowband signals as an extension of the Unitary ESPRIT for ULA [29]. The 1-D BMP algorithm was developed in [92] and [93] to estimate the DOA of the narrowband signals as an extension of DFT (Discrete Fourier Transform) beam-space ESPRIT for ULA [29]. Both of Single Invariance BMP (SBMP) and Multiple Invariance BMP (MBMP) algorithms were also investigated in [92]. In all MP algorithms, the most computationally intensive step is to estimate the signal subspace, which requires a Singular Value Decomposition (SVD) of a data matrix [90], [29]. By using the unitary matrix transformation (UMT) [94] in UMP algorithms and the DFT matrix transformation in BMP algorithms, the computational cost is reduced due to the conversion of the complex data matrix into real one, which is very efficient in real time implementations.

In this chapter, different realizations of the 1-D MP algorithms to estimate the TDOA are implemented based on the physical OFDM system parameters. The performance of these algorithms will be investigated and compared in the measurement chapter; the best will be used for wireless indoor positioning. The key element of our work is to use the preamble of the OFDM frame to measure the channel state for additional purposes to the demodulation of the data portion of the PPDU, which contains the training fields [50], [49], [48]. The complexity of using 20, 40, 80, and 160 MHz BWs of 802.11ac are investigated. The accuracy and stability of various MP algorithms with various 802.11ac BWs will be presented in the measurement chapter. A number of repeated training sequences are available in the OFDM frame, and the BSs within the service could be equipped with a number of antennas. Hence, the principle of using diversity techniques such as frequency diversity, time diversity, and space diversity is presented by using BWs partitioning, multiple OFDM training symbols, and a number of antennas, respectively.

Chapter 3. System Model and Time Delay Estimation using 1-D Matrix Pencil Algorithms

The various MP algorithms can use a single or multiple snapshots of the CFRs to estimate time delays, and then TDOA in the presence of multipath coherent signals. The results will be presented in Chapter 6.

The outline of this chapter is as follows: system model and OFDM system sensors principle are discussed in Section 3.1 to 3.3. A comparison between MP and other super-resolution algorithms is presented in Section 3.4. The enhanced 1-D MP, 1-D UMP, and 1-D BMP algorithms are presented in Sections 3.5, 3.6, and 3.7, respectively. The derivations of matrix pencil equations of these algorithms can be obtained from the relevant text of each algorithm. We will focus here on the main steps and our modifications to estimate the TDOA observations. The various 1-D MP algorithms will be enhanced to include the principle of diversity techniques in Section 3.8. In Section 3.9, computational complexity is presented to show the complexity of various 1-D MP algorithms.

3.1 OFDM Signal Model

Consider an OFDM system which consists of N_{FFT} subcarriers, where a number of useful subcarriers at the central spectrum $N_u + 1$ are used for transmission and the other subcarriers at both edges form the guard bands. The guard bands enable us to choose an appropriate analog transmission filter to limit the periodic spectrum of the discrete time signal at the output of the IFFT [95], [96]. Each transmission subcarrier is modulated by a pilot, data, or a null symbol $X_{i,k}$, where i represents the OFDM symbol number and k represents the subcarrier number. The IFFT of order N_{FFT} is usually used for modulation in OFDM transmitters, where N_{FFT} is a power of 2. A guard interval is also added as a cyclic prefix (CP) for every OFDM symbol to avoid intersymbol interference (ISI) caused by multipath fading channels. As long as the guard interval is larger than the maximum delay spread of the wireless channel, the ISI can be eliminated. As a result, the output baseband signal of the OFDM transmitter is represented:

$$x(t) = \sum_{i=-\infty}^{\infty} \sum_{k=-N_u/2}^{N_u/2} X_{i,k} e^{j\omega_k(t-T_g-iT_s)} \; ; iT_s \leq t < (i+1)T_s \qquad (3.1)$$

where $\omega_k = 2\pi k/T_u$, $1/T_u = \Delta f$ is the OFDM subcarrier spacing, T_g is the guard interval length, T_s is the duration of a whole OFDM symbol including the guard interval ($T_s = T_u + T_g$), and the sampling rate is $1/T = N_{FFT} \times \Delta f$.

At the receiver side, let us assume that the guard interval duration is longer than the channel maximum excess delay, the channel is quasi-stationary (i.e., the channel does not change within one OFDM symbol duration but varies from symbol to symbol), and the synchronization is done using the training sequence. From (2.1), the input signal at the receiver is

$$r(t) = \sum_{l=1}^{L} \alpha_l \cdot x(t - \tau_l) e^{j(\omega_c + \omega_l^D)(t-\tau_l)} + n(t) \qquad (3.2)$$

where ω_c and ω_l^D are the phase velocity of carrier and Doppler shift of the lth path, and $n(t)$ is the additive white Gaussian noise (AWGN) with mean zero and variance σ^2. From (3.1) and (3.2), we get

$$r_i(t) = \sum_{l=1}^{L} \sum_{k=-N_u/2}^{N_u/2} X_{i,k} e^{j\omega_k(t-\tau_l-T_g-iT_s)} \cdot \alpha_{i,l} \cdot e^{j(\omega_c+\omega_l^D)(t-\tau_l)} + n(t) \qquad (3.3)$$

where $iT_s \leq t < (i+1)T_s$. The RF component $\exp(j\omega_c t)$ is down converted at the receiver, then the multipath OFDM signal in the baseband channel is remained as

$$r_i(t) = \sum_{l=1}^{L} \sum_{k=-N_u/2}^{N_u/2} X_{i,k} e^{j\omega_k(t-\tau_l-T_g-iT_s)} \cdot \alpha_{i,l} \cdot e^{-j(\omega_c+\omega_l^D)\tau_l+\omega_l^D t} + n(t). \quad (3.4)$$

For packet detection and timing synchronization, the cross-correlation between the received samples and the training sequence, $s_{preamble}$, is used as [54], [97]

$$\Lambda(n) = \left| \sum_{u=0}^{N_{FFT}-1} r(u+n) \cdot s^*_{preamble}(u) \right|^2 \quad (3.5)$$

where $\{r(u)\}$ are the received samples. For robust timing acquisition, square low detection is used. In order also to avoid the expected variance of the incoming signal power, the cross-correlation needs to be normalized by a moving sum of the received signal power according to [98]

$$\Xi(n) = \Lambda(n)/\Upsilon(n) \quad (3.6)$$

where $\Upsilon(n) = \sum_{u=0}^{N_{FFT}-1} |r(u+n)|^2$. Once the FFT window has been adjusted, the kth subcarrier output during the ith OFDM symbol can be represented by

$$R_{i,k} = X_{i,k} \cdot \sum_{l=1}^{L} e^{-j2\pi k \Delta f \tau_l} \cdot A_{i,l} + w_{i,k}; \quad -N_u/2 \leq k \leq N_u/2 \quad (3.7)$$

where $w_{i,k}$ is the AWGN at the kth subcarrier, and $A_{i,l}$ is the channel gain of the lth path during the ith OFDM symbol as, $A_{i,l} = \alpha_{i,l} \cdot e^{-j(\omega_c+\omega_l^D)\tau_l}$. It has been assumed that the Doppler shifts $f_l^D = f_c v_l/c$ are much smaller than the subcarrier spacing ($f_l^D << \Delta f \Rightarrow \omega_l^D/2\pi\Delta f \approx 0$) and time delays are smaller than the guard interval ($\tau_l \leq T_g$) for all paths.

3.2 The symmetry of OFDM Time Delay and DOA Estimation Problems

To show the symmetry between the OFDM time delay and the DOA estimation problems, let us consider a ULA of M sensors as shown in Fig. 2.2 in Chapter 2. The distance between array elements is ρ. If L narrowband signals with carrier frequency f_c arrive at the input of this array, the measured data at the feeding point of the omni-directional antenna m is expressed [1]

$$y_m(n) = \sum_{i=1}^{L} \alpha_i e^{-j2\pi f_c \tau_m(\theta_i)} S_i(n) + n_m(n) \quad (3.8)$$

where α_i and θ_i are the complex gain and the DOA of the ith incoming signal, $S_i(n)$ is the ith signal at time instant n, $\tau_m(\theta_i) = m\rho\sin\theta_i/c$, and $n_m(n)$ is the AWGN at the mth sensor. The indexes of subsequent OFDM symbols in (3.7) are equivalent to the discrete time snapshots in (3.8). It can also be noted that the subcarrier spacing Δf between pilot symbols in OFDM symbol is proportional to the carrier frequency f_c, and the time delay τ_l of the lth path in a multipath channel is equivalent to the time delay that the plane wave impinging from direction θ_l needs to span the different distance between antenna m and the reference antenna in the antenna array [1]. Fig. 3.1 presents the

Chapter 3. System Model and Time Delay Estimation using 1-D Matrix Pencil Algorithms

OFDM signal in a multipath channel using a single antenna and the DOA problem using the ULA [1]. As a result, the problem of OFDM time delay estimation in a multipath channel is equivalent to the problem of DOA estimation using the ULA. Consequently, the known super-resolution algorithms for DOA estimation can be used for OFDM time delay estimation. Those algorithms can be classified as

- *Statistical algorithms* such as MUSIC, ROOT-MUSIC, and ESPRIT algorithms.
- *Non-statistical algorithms* such as MP algorithms.

3.3 OFDM System Sensors

In wireless communication systems, the training sequences are used for synchronization and estimating channel parameters. N_p pilot subcarriers per OFDM symbol are assumed. Pilot positions are $X_{i,k} = S_{i,m}$, where $m = 0, \ldots, N_P - 1$; the mth pilot is assigned to the kth subcarrier in an OFDM symbol. The least square estimate of the CFR can be obtained from (3.7) as

$$H_{i,k} = R_{i,k}/S_{i,m} = \sum_{l=1}^{L} z_l^k . A_{i,l} + w_{i,k}/S_{i,m} \qquad (3.9)$$

where $z_l = e^{-j2\pi \Delta f \tau_l}$. Therefore, the multipath channel poles (to be estimated) are

$$z_l = e^{-j\nu_l}; l = 1, \ldots, L \qquad (3.10)$$

where $\nu_l = 2\pi \Delta f \tau_l$. From (3.9), the estimated CFR can be modeled as the summation of the complex sinusoidal signals plus the complex white noise. Hence, if the time-domain variable τ_l is seen as a frequency-domain variable with L components, (3.9) can be seen as a harmonic signal model. Therefore, any spectral estimation method that is used for the harmonic signal model can be enhanced to use the CFR of the multipath radio channel to estimate the propagation time delays. In the following, the interested parts of the IEEE frame formats that are based on the OFDM technology are presented.

3.3.1 IEEE 802.11a Frame Format

The PLCP frame format includes the following fields: OFDM PLCP preamble, OFDM PLCP header, PLCP service data unit (PSDU), tail bits, and pad bits [50]. The PLCP preamble field is used for synchronization. It consists of 10 short symbols and two long symbols as shown in Fig. 3.2, where t_1 to t_{10} denote the short training symbols (STS), and T_1 and T_2 denote the long training symbols (LTS) [50]. The total length of the preamble is 16 μs. The PLCP preamble is followed by the signal field and data field. It should be noted that the STS field composes from 10 repetitions that are used for automatic gain control (AGC), diversity selection, timing acquisition, and coarse frequency acquisition in the receiver. The LTS field composes from two repetitions that are used for channel estimation and fine frequency acquisition in the receiver. It should also be noted that there are two LTS OFDM symbols that are transmitted to improve the channel estimation accuracy. The guard interval of LTS field is 1.6 μs, it is the half LTS sequence. The analysis bandwidth of IEEE 802.11a is 20 MHz.

3.3. OFDM System Sensors

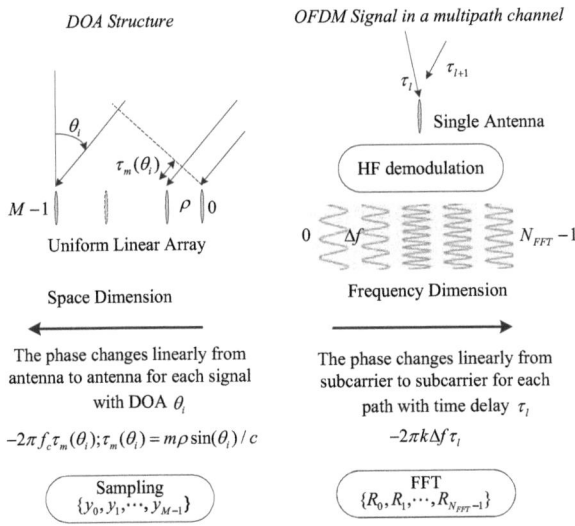

Figure 3.1: The symmetry of OFDM time delay and DOA estimation problems [1].

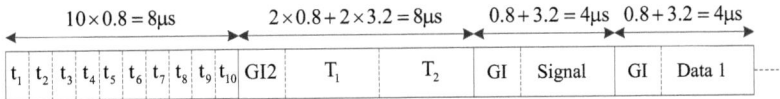

Figure 3.2: OFDM training structure [50].

Chapter 3. System Model and Time Delay Estimation using 1-D Matrix Pencil Algorithms

The short OFDM training symbol consists of 12 subcarriers, only every fourth subcarrier is occupied and the others are zeros to generate four repetitions in the time-domain. It is given by [50]

$$STS_{-26,26} = \sqrt{13/6} * \{0, 0, 1+1i, 0, 0, 0, -1-1i, 0, 0, 0, 1+1i, 0, 0, 0, -1-1i,$$
$$0, 0, 0, -1-1i, 0, 0, 0, 1+1i, 0, 0, 0, 0, 0, 0, -1-1i, 0, 0, 0, -1-1i,$$
$$0, 0, 0, 1+1i, 0, 0, 0, 1+1i, 0, 0, 0, 1+1i, 0, 0, 0, 1+1i, 0, 0\}. \tag{3.11}$$

To normalize the average power of the resulting OFDM symbol in (3.11), which utilizes 12 out of 52 subcarriers, it is multiplied by a factor of $\sqrt{13/6}$. The long OFDM training symbol consists of 53 subcarriers including a zero value at dc. It is given by [50]

$$LTS_{-26,26} = \{1,1,1,1,1,1,1,1,1,1,1,1,1,1,1,1,1,1,1,1,1,1,1,1,1,1,0$$
$$,1,1,1,1,1,1,1,1,1,1,1,1,1,1,1,1,1,1,1,1,1,1,1,1,1,1\}. \tag{3.12}$$

It is worth mentioning that all subcarriers in (3.12) are occupied while the subcarrier falling at dc ($0th$ subcarrier) is not used to avoid difficulties in digital to analog converter (D/A) and analog to digital converter (A/D) offsets and carrier feed through in the RF system [50]. Fig. 3.3 shows the spectrum of STS and LTS. Fig. 3.4 shows the transmitted preamble in the time-domain.

3.3.2 IEEE 802.11n and IEEE 802.11ac Frame Formats

Fig. 3.5 shows the high-throughput mixed (HT-mixed) and very high-throughput (VHT) PPDU formats used in our analysis for 802.11n [49] and 802.11ac [48], respectively. For synchronization, the low-throughput short training field (L-STF) and low-throughput Long Training Field (L-LTF) are used. In general, the physical parameters of 20 and 40 MHz BWs of both 802.11n and 802.11ac are equal. It should be noted that the L-STF, L-LTF, VHT-STF, and VHT-LTF portions of the preamble for 160 MHz VHT transmissions are constructed by repeating the 80 MHz counterparts twice in the frequency. As an example, the VHT-LTF is [48]

$$VHTLTF_{-250,250} = \{VHTLTF_{-122,122}, 0, 0, 0, 0, 0, 0, 0, 0, 0, 0, 0, VHTLTF_{-122,122}\}. \tag{3.13}$$

Equation (3.9) indicates that the OFDM time delay estimation problem can be seen as the DOA problem in antenna array-ULA [1], if the frequency distances between all pilot subcarriers are equal. The distance between pilot subcarriers within one OFDM symbol is limited. It needs that the change of phase between subsequent pilot symbols should not exceed 2π [26]. Therefore, we prefer using the LTS of 802.11a, the HT-LTF of 802.11n, and VHT-LTF of 802.11ac to achieve the highest range of estimated time delays, because they have the lowest frequency separation of $\Delta f = 312.5$ KHz.

Although a number of pilots in each OFDM data symbol are available, i.e. at 20 MHz BW, there are four pilots at $k = \{-21, -7, 7, 21\}$, they cannot be used due to the following reason. The change of phase between subsequent pilot symbols should not exceed 2π as

$$\Delta\phi = 2\pi\Delta f \tau_l a < 2\pi \Rightarrow a < \frac{T_u}{\tau_l}. \tag{3.14}$$

3.3. OFDM System Sensors

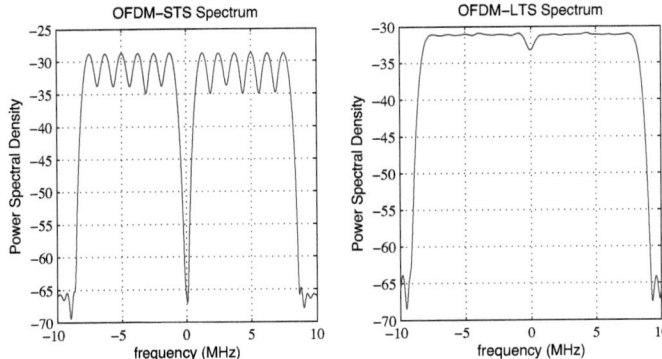

Figure 3.3: The spectrum of STS and LTS training sequences of 802.11a.

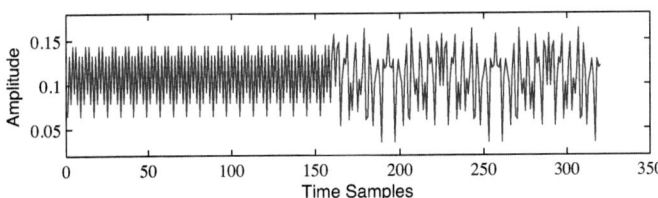

Figure 3.4: The transmitted preamble in the time-domain of 802.11a.

By taking into account that $\tau_l \leq T_g$, the separation between two neighboring pilots has to fulfill the following condition: $a \leq T_u/T_g = 4$. Consequently, the maximum separation distance between pilots in one OFDM symbol is $a = 4$. Therefor, the previous condition can be satisfied by using the long sequences of 802.11 OFDM frames.

The OFDM symbol structure of LTF has N_p symbols or sensors, and N_{dc} zero values at dc in the middle of LTF. Table 3.1 shows the LTF parameters of each bandwidth. The number of occupied subcarriers at 20 MHz BW of 802.11a and 802.11n is 52 and 56, respectively. These zero values at dc in the middle of LTF lead to an array discontinuity of the estimated CFR. Hence, an interpolation is used to mitigate the array discontinuity as we have proposed in [53] and [54]. The useful number of pilots is consequently, $N = N_p + N_{dc}$. The number of LTS per frame is 2 and the number of HT-LTFs per frame could be 1, 2, or 4, and the number of VHT-LTFs per frame could be 1, 2, 4, 6, or 8 that are necessary for the demodulation of the PPDU data or for channel estimation during a null data packet (NDP) [48], [49].

3.4 Super-Resolution Algorithms for Wireless Indoor Positioning

An effective wireless indoor positioning can be implemented using the opportune signals of IEEE 802.11 standards. The largest channel bandwidths of 802.11a and 802.11n are 20 and 40 MHz, respectively, while it is 160 MHz in 802.11ac. Therefore, super-resolution algorithms should be used for post processing to reduce bandwidth requirements as it has been explained in Section 2.4. They can provide a high resolution DOA estimation based on the eigen-decomposition of the covariance matrix such as MUSIC [99], ROOT-MUSIC [20], and ESPRIT [28], or the eigen-decomposition of the data matrix directly such as MP algorithm [34].

Based on the symmetry of the DOA and OFDM time delay estimation problems presented in Section 3.2, all subspace-based algorithms can be used to estimate the time delays of a multipath channel based on the estimated CFR using the preamble of the OFDM frame. In case of covariance matrix based techniques, a minimum of L independent CFR estimates of the same channel are required to estimate the time delays of L paths. The estimated CFRs must be obtained repeatedly, which is a time-consuming and wasteful process. In addition, the channel should not vary over L or more estimates [100], where the wireless channel can vary rapidly. Furthermore, additional spatial-smoothing methods should be applied to modify the covariance matrix to distinguish between correlated channel components to solve the rank condition problem of the covariance matrix [91], [100].

Table 3.1: The lengths of LTS and LTF regarding the 20 MHz BW of 802.11a and all BWs of 802.11ac.

Bandwidth (MHz)	20 (a)	20	40	80	160
FFT/IFFT order	64	64	128	256	512
N_p	52	56	114	242	484
N_{dc}	1	1	3	3	3+11+3

3.5. 1-D Matrix Pencil Algorithm

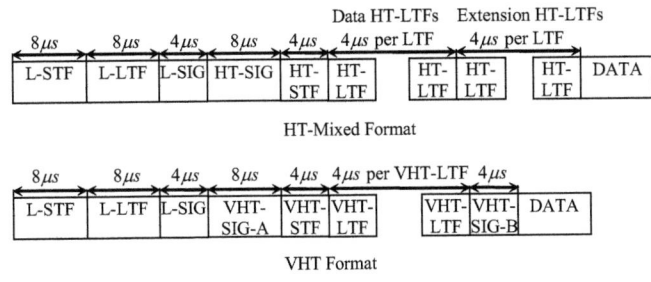

Figure 3.5: PPDU formats of 802.11n and 802.11ac.

On the other hand, the MP algorithm works with the data directly without forming a covariance matrix. The snapshot-by-snapshot analysis is used, hence, non-stationary environments can be handled easily. MP algorithm can find the DOA in the presence of multipath coherent signals without performing additional processing like spatial smoothing as required in the conventional covariance matrix based techniques [90]. Table 3.2 summarizes the comparison between the MP algorithm and the covariance matrix based algorithms. Therefore, the principle of MP algorithm is superior compared with other super-resolution algorithms such as ESPRIT and Root MUSIC. As a result, the principle of MP algorithms represents our concern in this work to develop for wireless indoor positioning.

3.5 1-D Matrix Pencil Algorithm

The MP algorithm was originally developed in order to estimate the poles of a system [34], [35], [101]. It is more efficient in computation, less restrictive about signal poles, and less sensitive to noise of signals with unknown damping factors than the polynomial method [34]. In the literature, the 1-D MP algorithm has been presented based on the uniform space between array elements using the ULA to estimate multiple frequencies or angle of arrivals of narrowband signals. In this work, the 1-D MP algorithms are used to estimate the time delays using the OFDM pilots, which are distributed equally in the frequency dimension. It has been published in [53] and [54].

The objective is to find the set $\{z_l\}$ presented in (3.10), which corresponds to the time delays. The rank of the estimated CFR vector $\mathbf{H}_{N \times 1}$ defined in (3.9) is one. To enhance the rank condition (the dimensionality of the signal subspace), a partition and stacking process should be used. By using the principle of the conventional MP [34], the Hankel matrix $\mathbf{Y}$ can be created by windowing the estimated CFR vector $\mathbf{H}$ as follows

$$\mathbf{Y} = \begin{bmatrix} H_0 & H_1 & \cdots & H_P \\ H_1 & H_2 & \cdots & H_{P+1} \\ \vdots & \vdots & \ddots & \vdots \\ H_{N-P-1} & H_{N-P} & \cdots & H_{N-1} \end{bmatrix}_{(N-P) \times (P+1)} \quad (3.15)$$

Chapter 3. System Model and Time Delay Estimation using 1-D Matrix Pencil Algorithms

Table 3.2: The comparison between the principle of MP algorithm and that of the covariance matrix based algorithms.

Description	Matrix Pencil	MUSIC and ESPRIT
First step	Works with the measured data directly.	Statistical methods by finding the covariance matrix.
To estimate L paths	Only a single channel estimate is used.	More than L independent channel estimates are used.
Channel Status	Channel must remain constant over a single estimate. Non-stationary environments can be handled easily.	Channel must remain constant over several estimates. Non-stationary environments cannot be handled easily.
In relatively rapidly varying channels	It can be applied.	They cannot be applied.
For channel correlated components	It can distinguish between them.	A preprocessing, called spatial smoothing, is necessary to distinguish between them.
In coherent multi-path scenarios	It can be applied.	They failed to produce a successful recovery.
Time and Complexity	Time saving and less complex. It can be applied for real time applications.	Time consuming and wasteful process.

where P is the pencil parameter which plays an important role in reducing noise sensitivity. From (3.15), the value of pencil parameter P should be determined to generate the Hankel matrix. In fact, P can be selected as, $L \leq P \leq N - L$, for noiseless data. For noisy data, the value of P should be selected to provide a balance between resolution, complexity, and stability of the algorithm. Different values of P have been used in the literature for algorithms that use only single snapshot. If the number of antennas is N, the value of P has been selected in [102], [103], [104] as $N/2, 3N/4, 3N/5$, respectively. However, it has been found in [34] that for efficient noise filtering in the conventional MP algorithm, the parameter P should be selected between $N/3$ and $2N/3$. We will investigate that for time delay estimation problem using OFDM systems in the measurement chapter.

For noiseless data, the two submatrices $\mathbf{Y}_1$ and $\mathbf{Y}_2$ of the matrix pencil can be defined in terms of $\mathbf{Y}$ by deleting the last and the first row, respectively, as

$$\mathbf{Y}_1 = \begin{bmatrix} H_0 & H_1 & \cdots & H_P \\ H_1 & H_2 & \cdots & H_{P+1} \\ \vdots & \vdots & \ddots & \vdots \\ H_{N-P-2} & H_{N-P-1} & \cdots & H_{N-2} \end{bmatrix}_{(N-P-1)\times(P+1)}, \qquad (3.16)$$

3.5. 1-D Matrix Pencil Algorithm

$$\mathbf{Y}_2 = \begin{bmatrix} H_1 & H_2 & \cdots & H_{P+1} \\ H_2 & H_3 & \cdots & H_{P+2} \\ \vdots & \vdots & \ddots & \vdots \\ H_{N-P-1} & H_{N-P} & \cdots & H_{N-1} \end{bmatrix}_{(N-P-1) \times (P+1)} \quad (3.17)$$

From (3.9), in case of noiseless data, the matrices of (3.16) and (3.17) can be written as [35]

$$\mathbf{Y}_1 = \mathbf{Z}_1 \mathbf{A} \mathbf{Z}_2 \quad (3.18)$$

$$\mathbf{Y}_2 = \mathbf{Z}_1 \mathbf{A} \mathbf{Z}_d \mathbf{Z}_2 \quad (3.19)$$

where

$$\mathbf{Z}_1 = \begin{bmatrix} 1 & 1 & \cdots & 1 \\ z_1 & z_2 & \cdots & z_L \\ \vdots & \vdots & \ddots & \vdots \\ z_1^{N-P-2} & z_2^{N-P-2} & \cdots & z_L^{N-P-2} \end{bmatrix}_{(N-P-1) \times L}, \quad (3.20)$$

$$\mathbf{Z}_2 = \begin{bmatrix} 1 & z_1 & \cdots & z_1^P \\ 1 & z_2 & \cdots & z_2^P \\ \vdots & \vdots & \ddots & \vdots \\ 1 & z_L & \cdots & z_L^P \end{bmatrix}_{L \times (P+1)}, \quad (3.21)$$

$$\mathbf{Z}_d = \text{diag}\left[z_1, z_2, \ldots, z_L\right], \quad (3.22)$$

$$\mathbf{A} = \text{diag}[\alpha_1 e^{-j\omega_c \tau_1}, \ldots, \alpha_L e^{-j\omega_c \tau_L}]. \quad (3.23)$$

Then, the matrix pencil equation is

$$\mathbf{Y}_2 - \eta \mathbf{Y}_1 = \mathbf{Z}_1 \mathbf{A} [\mathbf{Z}_d - \eta \mathbf{I}] \mathbf{Z}_2 \quad (3.24)$$

where $\mathbf{I}$ is the identity matrix. The rank of $\mathbf{Y}_2 - \eta \mathbf{Y}_1$ is the number of effective paths L while the pencil value is $L \leq P \leq N - L$ [35], [101]. However, if $\eta = z_l; l = 1, \ldots, L$, the rank of the pencil matrix reduces by one. Therefore, the estimates of multipath channel poles $\{z_l\}$ are the generalized eigenvalues of the matrix pair $\{\mathbf{Y}_2, \mathbf{Y}_1\}$, which can be computed as

$$\mathbf{Y}_2 - \eta \mathbf{Y}_1 \Rightarrow \mathbf{Y}_1^\dagger \mathbf{Y}_2 - \eta \mathbf{Y}_1^\dagger \mathbf{Y}_1 \Rightarrow \mathbf{Y}_1^\dagger \mathbf{Y}_2 - \eta \mathbf{I} \quad (3.25)$$

where the superscript $\dagger$ denotes the Moore-Penrose pseudo-inverse. It is defined for $\mathbf{Y}_1$ as

$$\mathbf{Y}_1^\dagger = \left(\mathbf{Y}_1^H \mathbf{Y}_1\right)^{-1} \mathbf{Y}_1^H \quad (3.26)$$

where the superscript H denotes the conjugate transpose.

Summary: The enhanced 1-D MP algorithm for channel profile parameters estimation and then TDOA estimation will be summarized in the following:

Step1: From the estimated CFR, find the Hankel matrix $\mathbf{Y}$ as in (3.15). To increase the number of snapshots inherently, an extended matrix $\mathbf{Y}_{ex}$ could be defined in forward and backward:

$$\mathbf{Y}_{ex} = [\mathbf{Y} : \underset{N-P}{\Pi} \mathbf{Y}^* \underset{P+1}{\Pi}] \quad (3.27)$$

Chapter 3. System Model and Time Delay Estimation using 1-D Matrix Pencil Algorithms

where : means matrix partitioning, and Π is the exchange matrix that reverses the ordering of the rows; it has been defined in (A.3) of appendix A. The extended matrix can also be generated:

$$\mathbf{Y}_{ex} = [\mathbf{Y} : \underset{N-P}{\Pi} \mathbf{Y}^*]. \quad (3.28)$$

If we have a vector of 9 samples, Table 3.3 shows the samples order using (3.27) and (3.28). Both of them give the same results. In this work, the form of (3.27) will be used.
Step2: The SVD is used for noisy data to reduce part of the noise effect and to get the signal subspace. The matrix $\mathbf{Y}$ of (3.15) (or $\mathbf{Y}_{ex}$ of (3.27)) can then be decomposed as follows [83]

$$\mathbf{Y} = \mathbf{U}\Sigma\mathbf{V}^H \quad (3.29)$$

where $\mathbf{U} = \{\mathbf{u}_1, \cdots, \mathbf{u}_{N-P}\}$ is $(N-P \times N-P)$ unitary matrix composed of eigenvectors of $\mathbf{YY}^H$, $\mathbf{V} = \{\mathbf{v}_1, \cdots, \mathbf{v}_{P+1}\}$ is $(P+1 \times P+1)$ unitary matrix composed of eigenvectors of $\mathbf{Y}^H\mathbf{Y}$, and $\Sigma = \sqrt{eig(\mathbf{Y}^H\mathbf{Y})}$ is a diagonal matrix with the singular values (SVs) of $\mathbf{Y}$ being, $\eta_1 \geq \eta_2 \geq \cdots \geq \eta_B$, where $B = min(N-P, P+1)$. The B-dimensional subspace of the signal vector can be split into two orthogonal subspaces: the signal subspace and the noise subspace. Hence, the matrix Σ can be split into two submatrices Σ_s and Σ_n. Σ_s is $L \times L$ diagonal matrix with the L largest SVs which characterize the signal subspace as, $\eta_k = \eta_k^s + \sigma_k^2$, where $k = 1, \ldots, L$. Σ_n is $(B-L) \times (B-L)$ diagonal matrix with the $(B-L)$ smallest SVs which characterize the additive noise as, $\eta_k = \sigma_k^2$, where $k = L+1, \ldots, B$. If the forward-backward matrix $\mathbf{Y}_{ex}$ of (3.27) is used in (3.29), $B = min(N-P, 2(P+1))$; it will be called the 1-D MP-Ex algorithm.
Step3: Based on the Information Theoretic Criteria (ITC), the Minimum Descriptive Length (MDL) is used to estimate the signal subspace dimension $\hat{L}$ by eliminating the noise components [105]. The log-likelihood function of [105] and the modified penalty function of [106] are used, in order to bias the model over estimation, according to

$$MDL(k) = -(B-k)P \cdot \log \left\{ \frac{\prod_{i=k+1}^{B} \eta_i^{1/(B-k)}}{\frac{1}{B-k}\sum_{i=k+1}^{B} \eta_i} \right\} + \frac{1}{4}k(2B-k)\log(P) + k \quad (3.30)$$

where $\eta_i, 0 \leq i \leq B-1$, are the SVs in a descending order. It is worth mentioning that the modified MDL does not require any subjective threshold such as the conventional hypothesis [105]. The estimated number of arrived signals $\hat{L}$ is determined as the value of $k \in \{0, 1, \ldots, B-1\}$ for which the modified MDL is minimized. From (3.30), the term in the brackets is the ratio of the geometric mean to the arithmetic mean of the smallest $B-k$ SVs. To simplify (3.30), both terms can be divided by P.

The principle of MDL criterion of detecting the number of arrival signals by using the SVs is that the SVs corresponding to noise subspace are equal to the noise power and those noise SVs have roots in Gaussian distribution. The cost function has been assumed based on Gaussian distribution using information theory. This cost function estimates the threshold which divides the set of SVs into that of signal and that of noise.
Step4: The SVD of $\mathbf{Y}$ in (3.15) or $\mathbf{Y}_{ex}$ in (3.27) can then be decomposed as

$$\mathbf{Y} = \mathbf{U}\Sigma\mathbf{V}^H = \mathbf{Y}_s + \mathbf{Y}_n = \mathbf{U}_s\Sigma_s\mathbf{V}_s^H + \mathbf{U}_n\Sigma_n\mathbf{V}_n^H \quad (3.31)$$

3.5. 1-D Matrix Pencil Algorithm

Table 3.3: The distribution of samples inside both possible generations of the extended matrix using the exchange matrix.

$\mathbf{Y}_{ex} = [\mathbf{Y} : \mathbf{\Pi}\,\mathbf{Y}^*\mathbf{\Pi}]$										
1	2	3	4	5	6	*9*	8	7	6	5	4
2	3	4	5	6	7	8	7	6	5	4	3
3	4	5	6	7	8	7	6	5	4	3	2
4	5	6	7	8	*9*	6	5	4	3	2	*1*

$\mathbf{Y}_{ex} = [\mathbf{Y} : \mathbf{\Pi}\,\mathbf{Y}^*]$										
1	2	3	4	5	6	4	5	6	7	8	*9*
2	3	4	5	6	7	3	4	5	6	7	8
3	4	5	6	7	8	2	3	4	5	6	7
4	5	6	7	8	*9*	*1*	2	3	4	5	6

where $\mathbf{U}_s$ and $\mathbf{V}_s$ are the submatrices of $\mathbf{U}$ and $\mathbf{V}$ corresponding to the largest $\hat{L}$ singular values of $\mathbf{\Sigma}_s$ and span the signal subspace while $\mathbf{U}_n$ and $\mathbf{V}_n$ are corresponding to the singular values of $\mathbf{\Sigma}_n$ and span the noise subspace. From (3.31), if the number of columns in $\mathbf{U}_n$ increases, which represents the dimension of the noise subspace, more noise components can be absorbed into the noise subspace, where a larger noise subspace implies high estimation accuracy [36]. Therefore, the term $N - P$ should be large enough to include the signal subspace and the noise subspace dimensions. However, $N - P$ should be smaller than $P + 1$ to reduce the overall computational complexity, as it will be explained in Section 3.9.

Step5: The filtered matrices $\mathbf{Y}_{s1}$ and $\mathbf{Y}_{s2}$ can be obtained from $\mathbf{Y}_s$ instead of $\mathbf{Y}$ in the conventional 1-D MP algorithm, or $\mathbf{Y}_{ex}$ in the 1-D MP-Ex algorithm. $\mathbf{Y}_{s1}$ and $\mathbf{Y}_{s2}$ can be defined as, $\mathbf{Y}_{s1} = \mathbf{U}_{s1}\mathbf{\Sigma}_s\mathbf{V}_s^H$, $\mathbf{Y}_{s2} = \mathbf{U}_{s2}\mathbf{\Sigma}_s\mathbf{V}_s^H$, where $\mathbf{U}_{s1}$ and $\mathbf{U}_{s2}$ are obtained from $\mathbf{U}_s$ by deleting the last and the first row, respectively. The desired z_l's can be obtained as the eigenvalues of $\mathbf{U}_{s1}^\dagger \mathbf{U}_{s2}$, which represents a generalized eigenvalue problem of dimension $\hat{L} \times \hat{L}$. In case of using (3.15), the two submatrices $\mathbf{Y}_{s1}$ and $\mathbf{Y}_{s2}$ can also be defined in terms of $\mathbf{Y}_s$ by deleting the last and the first column, respectively. The generalized eigenvalue problem is then that of $\mathbf{V}_{s2}^H[\mathbf{V}_{s1}^H]^\dagger$, where $\mathbf{V}_{s1}$ and $\mathbf{V}_{s2}$ are obtained from $\mathbf{V}_s$ by deleting the last and the first row, respectively [35].

Step6: The lth time delay can then be calculated as

$$z_l = e^{-j2\pi\Delta f \hat{\tau}_l} \Rightarrow \hat{\tau}_l = \frac{\arg(z_l^*)}{2\pi\Delta f}; l = 1, ..., \hat{L} \qquad (3.32)$$

where $\arg(z_l^*)$ denotes the phase of z_l^*, and the superscript * denotes the complex conjugate.

Once the $\hat{L}$ effective poles $\mathbf{z} = [z_1, ..., z_{\hat{L}}]^T$ are estimated, the channel gains $\mathbf{A}_0 = [\alpha_1 e^{-j\omega_c \tau_1}, ..., \alpha_{\hat{L}} e^{-j\omega_c \tau_{\hat{L}}}]^T$ can be obtained by solving a least square problem, where the superscript ^{T} denotes the transpose. To make notations simpler, we rewrite (3.9) in a matrix form:

$$\mathbf{H} = \mathbf{Z}\mathbf{A}_0 \qquad (3.33)$$

where

$$\mathbf{Z} = \begin{bmatrix} z_1^{-\frac{(N-1)}{2}} & z_2^{-\frac{(N-1)}{2}} & \cdots & z_{\hat{L}}^{-\frac{(N-1)}{2}} \\ \vdots & \vdots & \ddots & \vdots \\ 1 & 1 & 1 & 1 \\ \vdots & \vdots & \ddots & \vdots \\ z_1^{\frac{(N-1)}{2}} & z_2^{\frac{(N-1)}{2}} & \cdots & z_{\hat{L}}^{\frac{(N-1)}{2}} \end{bmatrix}_{N \times \hat{L}} \tag{3.34}$$

The channel complex gains can then be calculated:

$$\mathbf{A}_0 = \mathbf{Z}^\dagger \mathbf{H} \tag{3.35}$$

3.6 1-D Unitary Matrix Pencil Algorithm

The UMT was applied successfully in [90] for 1-D UMP algorithm, in which the complexity of the computation is reduced by using real computations. In this section, we will develop the principle of the computationally efficient 1-D UMP algorithm for wireless positioning using the wideband orthogonal multi-carrier signals represented by OFDM to estimate the time delays of a wireless multipath channel. The results have been published in [53] and [54], as we will see in the measurement chapter. To do that, three theorems of the UMT are required presented in appendix A, where their proofs can be found in [90].

By using the centro-symmetry of the OFDM (corresponding to that in the antenna arrays) or any complex matrix could be written in a centro-hermitian matrix form, the centro-Hermitian matrices $\mathbf{Y}$ and $\mathbf{Y}_{ex}$ defined in (3.15) and (3.27) can be converted to a real matrix $\mathbf{Y}_{\text{Re}}$ using theorem 3 in appendix A as

$$\mathbf{Y}_{\text{Re}} = \mathbf{Q}_{K_1}^H \mathbf{Y} \mathbf{Q}_{K_2} \quad \text{or} \quad \mathbf{Y}_{\text{Re}} = \mathbf{Q}_{K_1}^H \mathbf{Y}_{ex} \mathbf{Q}_{K_2} \tag{3.36}$$

where $\mathbf{Q}_{K_1}$ and $\mathbf{Q}_{K_2}$ are unitary matrices whose columns are conjugate symmetric and have the sparse structure as

$$\mathbf{Q}_{a(even)} = \frac{1}{\sqrt{2}} \begin{bmatrix} \mathbf{I} & j\mathbf{I} \\ \mathbf{\Pi} & -j\mathbf{\Pi} \end{bmatrix}, \tag{3.37}$$

$$\mathbf{Q}_{b(odd)} = \frac{1}{\sqrt{2}} \begin{bmatrix} \mathbf{I} & \mathbf{0}^T & j\mathbf{I} \\ 0 & \sqrt{2} & 0 \\ \mathbf{\Pi} & \mathbf{0}^T & -j\mathbf{\Pi} \end{bmatrix}. \tag{3.38}$$

For $\mathbf{Q}_a$, the identity matrix $\mathbf{I}$ and the exchange matrix $\mathbf{\Pi}$ have the dimension of $a/2$, for $\mathbf{Q}_b$, $\mathbf{I}$ and $\mathbf{\Pi}$ have the dimension of $(b-1)/2$, and $\mathbf{0}$ is a $(b-1)/2$ zero row vector.

After that some selection matrices are used to get the matrix pencil equation. The selection matrices $\mathbf{J}_1$ and $\mathbf{J}_2$ are used to select the rows of the real matrix $\mathbf{Y}_{\text{Re}}$ in order to write the matrix pencil equation. They are given by

$$\mathbf{J}_1 = [\mathbf{I}_{N-P-1} : \mathbf{0}_{(N-P-1)\times 1}], \tag{3.39}$$

$$\mathbf{J}_2 = [\mathbf{0}_{(N-P-1)\times 1} : \mathbf{I}_{N-P-1}]. \tag{3.40}$$

3.6. 1-D Unitary Matrix Pencil Algorithm

In order to estimate the set of poles $\{z_l\}$, the matrix pencil equation can be written as

$$\mathbf{J}_2\mathbf{Y} = z\mathbf{J}_1\mathbf{Y} \quad \text{or} \quad \mathbf{J}_2\mathbf{Y}_{ex} = z\mathbf{J}_1\mathbf{Y}_{ex}. \tag{3.41}$$

Equation (3.41) can be written in terms of $\mathbf{Y}_{ex}$ as

$$\mathbf{Q}^H\mathbf{J}_2\mathbf{Q}\mathbf{Q}^H\mathbf{Y}_{ex}\mathbf{Q} = z\mathbf{Q}^H\mathbf{J}_1\mathbf{Q}\mathbf{Q}^H\mathbf{Y}_{ex}\mathbf{Q} \tag{3.42}$$

where the unitary matrices $\mathbf{Q}$ have proper dimensions. From (3.36), (3.42) becomes

$$\mathbf{Q}^H\mathbf{J}_2\mathbf{Q}\mathbf{Y}_{\text{Re}} = z\mathbf{Q}^H\mathbf{J}_1\mathbf{Q}\mathbf{Y}_{\text{Re}}. \tag{3.43}$$

By using the next properties $\Pi\Pi = \mathbf{I}$, $\Pi\mathbf{Q} = \mathbf{Q}^*$, $\mathbf{Q}^H\Pi = \mathbf{Q}^T$, and $\Pi\mathbf{J}_2\Pi = \mathbf{J}_1$, where the sizes of the corresponding matrices Π and $\mathbf{Q}$ are appropriate, it has been proved in [90] that

$$\mathbf{Q}^H\mathbf{J}_2\mathbf{Q} = (\mathbf{Q}^H\mathbf{J}_1\mathbf{Q})^*. \tag{3.44}$$

Therefore, (3.43) can be written as

$$(\mathbf{Q}^H\mathbf{J}_1\mathbf{Q})^*\mathbf{Y}_{\text{Re}} = z\mathbf{Q}^H\mathbf{J}_1\mathbf{Q}\mathbf{Y}_{\text{Re}}. \tag{3.45}$$

As it is described in the previous, $\mathbf{z} = \{z_l; l = 1, ..., \hat{L}\}$ are the rank reducing numbers of the matrix pencil equation, hence, (3.45) can be written as

$$(\mathbf{Q}^H\mathbf{J}_1\mathbf{Q})^*\mathbf{Y}_{\text{Re}} = e^{-j\nu_l}\mathbf{Q}^H\mathbf{J}_1\mathbf{Q}\mathbf{Y}_{\text{Re}} \tag{3.46}$$

where $\nu_l = 2\pi\Delta f \hat{\tau}_l$ defined in (3.10). By using the definition of the tangent function with some mathematical manipulations, it can be shown that

$$\tan(\nu_l/2)\mathbf{K}_{\text{Re}}\mathbf{Y}_{\text{Re}} = \mathbf{K}_{\text{Im}}\mathbf{Y}_{\text{Re}} \tag{3.47}$$

where

$$\mathbf{K}_{\text{Re}} = \text{Re}(\mathbf{Q}^H_{K_1}\mathbf{J}_1\mathbf{Q}_{K_2}) \quad \mathbf{K}_{\text{Im}} = \text{Im}(\mathbf{Q}^H_{K_1}\mathbf{J}_1\mathbf{Q}_{K_2}). \tag{3.48}$$

By using the same principle of 1-D MP, the matrix $\mathbf{Y}_{\text{Re}}$ can be replaced by its signal subspace $\mathbf{U}_s$ in order to reduce the effect of noise. Therefore, the matrix pencil equation is

$$\mathbf{K}_{\text{Re}}\mathbf{U}_s\mathbf{\Psi}_\nu = \mathbf{K}_{\text{Im}}\mathbf{U}_s. \tag{3.49}$$

The generalized eigenvalues of pencil pair $(\mathbf{K}_{\text{Im}}\mathbf{U}_s, \mathbf{K}_{\text{Re}}\mathbf{U}_s)$ can be calculated as the eigenvalues of

$$\mathbf{\Psi}_\nu = [\mathbf{K}_{\text{Re}}\mathbf{U}_s]^\dagger \mathbf{K}_{\text{Im}}\mathbf{U}_s \tag{3.50}$$

to get $\mathbf{Z}_d = \text{diag}\{\Lambda_l = \tan(\nu_l/2); l = 1, \dots, \hat{L}\}$. Then, calculate the time delays according to $\hat{\tau}_l = 1/\pi\Delta f \times \tan^{-1}(\Lambda_l)$, where $l = 1, \dots, \hat{L}$.

Summary: The enhanced 1-D UMP algorithm for channel profile parameters estimation and then TDOA estimation will be summarized in the following:

Step1: The rank of the CFR vector $\mathbf{H}_{N\times 1}$ defined in (3.9) is one. To enhance the rank condition, a partition and stacking process should be used. Therefore, the Hankel matrix $\mathbf{Y}$ is defined as in (3.15) or the extended matrix $\mathbf{Y}_{ex}$ as in (3.27).

Step2: Calculate the real data matrix $\mathbf{Y}_{\text{Re}}$, using the centro-symmetry of the OFDM, which could be written in a centro-hermitian matrix form, by either using $\mathbf{Y}$ of (3.15) (will be called UMP algorithm) or by using $\mathbf{Y}_{ex}$ of (3.27) (will be called UMP-Ex algorithm) as, $\mathbf{Y}_{Re} = \mathbf{Q}_{K_1}^H \mathbf{Y}_{ex} \mathbf{Q}_{K_2}$, where $\mathbf{Q}_{K_1}$ and $\mathbf{Q}_{K_2}$ are unitary matrices given in (3.37) and (3.38).

Step3: Perform the SVD on $\mathbf{Y}_{Re}$ as in (3.31) to get the signal subspace, where $\mathbf{Y}_{Re}$ can be decomposed as, $\mathbf{Y}_{Re} = \mathbf{U}\mathbf{\Sigma}\mathbf{V}^H = \mathbf{U}_s\mathbf{\Sigma}_s\mathbf{V}_s^H + \mathbf{U}_n\mathbf{\Sigma}_n\mathbf{V}_n^H$. Then, estimate the number of effective paths $\hat{L}$ using the modified MDL criterion as in (3.30).

Step4: Determine $\mathbf{U}_s$ as the left singular vectors of $\mathbf{U}$, which span the signal subspace and correspond to the largest $\hat{L}$ singular values of $\mathbf{Y}_{Re}$.

Step5: Calculate the selection matrices $\mathbf{K}_{Re}$ and $\mathbf{K}_{Im}$ to extract the $\hat{L}$ poles $\{z_l\}$ just once and store them as constant matrices according to $\mathbf{K}_{Re} = \text{Re}(\mathbf{Q}_{K_1}^H \mathbf{J}_1 \mathbf{Q}_{K_2})$ and $\mathbf{K}_{Im} = \text{Im}(\mathbf{Q}_{K_1}^H \mathbf{J}_1 \mathbf{Q}_{K_2})$, where $\mathbf{J}_1$ is a selection matrix constructed from an identity matrix and a zero vector according to $\mathbf{J}_1 = [\mathbf{I}_{N-P-1} : \mathbf{0}_{(N-P-1)\times 1}]$.

Step6: Finally, the desired $z_l's$ can be obtained by computing the generalized eigenvalue problem of $[\mathbf{K}_{Re}\mathbf{U}_s]^\dagger \mathbf{K}_{Im}\mathbf{U}_s$ to get $\{\Lambda_l = \tan(\nu_l/2); l = 1, \ldots, \hat{L}\}$. The required time delays can then be calculated according to $\hat{\tau}_l = 1/\pi\Delta f \times \tan^{-1}(\Lambda_l)$, where $l = 1, \ldots, \hat{L}$.

Such as in 1-D MP, once the $\hat{L}$ effective poles $\mathbf{z} = [z_1, \ldots, z_{\hat{L}}]^T$ are estimated, the channel gains $\mathbf{A}_0 = [\alpha_1 e^{-j\omega_c \tau_1}, \ldots, \alpha_{\hat{L}} e^{-j\omega_c \tau_{\hat{L}}}]^T$ can be obtained by solving a least square problem as in (3.35).

3.7 1-D Beam-space Matrix Pencil Algorithm

The DFT was applied successfully in [92] for 1-D BMP algorithm to estimate the DOA of the narrowband signals, in which the complexity of the computation is reduced by using real valued computations. In this section, the principle of 1-D BMP has also been developed for wireless positioning systems to estimate time delays of a wireless multipath channel. The results have been published in [53] and [54], as we will see in the measurement chapter.

Let us assume that $\mathbf{F}_C$ is the DFT matrix of dimensions $C \times C$, where $C = N - P$ to simplify the notations. The conjugate centro-symmetrized version of the mth row of $\mathbf{F}_C$ can be written as

$$\mathbf{f}_m^H = e^{j\frac{C-1}{2}m\frac{2\pi}{C}} \times [1, e^{-jm\frac{2\pi}{C}}, e^{-j2m\frac{2\pi}{C}}, \ldots, e^{-j(C-1)m\frac{2\pi}{C}}] \qquad (3.51)$$

where the row vector $\mathbf{f}_m^H$ represents the DFT beam steered at spatial frequency, $\nu = m \times 2\pi/C$.

To get the matrix pencil equation of BMP, let us re-describe the necessary derivation of [93] and [29] for wireless positioning. The noiseless Hankel matrix of (3.15) can be represented as

$$\mathbf{Y} = \mathbf{Z}_1 \mathbf{A} \mathbf{Z}_2 \Rightarrow \mathbf{Y} = \mathbf{Z}_1' \mathbf{Z}_d^{\frac{C-1}{2}} \mathbf{A} \mathbf{Z}_2 \qquad (3.52)$$

where $\mathbf{Z}_2$, $\mathbf{Z}_d$, and $\mathbf{A}$ are defined in (3.21), (3.22), and (3.23), respectively, $\mathbf{Z}_1$ is similar

3.7. 1-D Beam-space Matrix Pencil Algorithm

to (3.20) without deleting the last row, and

$$\mathbf{Z}_1' = \begin{bmatrix} z_1^{\frac{-(C-1)}{2}} & z_2^{\frac{-(C-1)}{2}} & \cdots & z_L^{\frac{-(C-1)}{2}} \\ z_1^{\frac{-(C-3)}{2}} & z_2^{\frac{-(C-3)}{2}} & \cdots & z_L^{\frac{-(C-3)}{2}} \\ \vdots & \vdots & \ddots & \vdots \\ z_1^{\frac{(C-3)}{2}} & z_2^{\frac{(C-3)}{2}} & \cdots & z_L^{\frac{(C-3)}{2}} \\ z_1^{\frac{(C-1)}{2}} & z_2^{\frac{(C-1)}{2}} & \cdots & z_L^{\frac{(C-1)}{2}} \end{bmatrix}_{C \times L}. \tag{3.53}$$

Let us assume $\nu_l = -2\pi\Delta f \tau_l$, hence, the multipath channel poles to be estimated are $\{z_l = e^{j\nu_l}\}$. Then, multiply (3.52) by the conjugate centro-symmetrized DFT matrix $\mathbf{F}_C^H$ whose rows given by (3.51) as follows

$$\mathbf{Y}_F = \mathbf{F}_C^H \mathbf{Z}_1' \mathbf{Z}_d^{\frac{C-1}{2}} \mathbf{A} \mathbf{Z}_2 = \mathbf{B} \mathbf{Z}_d^{\frac{C-1}{2}} \mathbf{A} \mathbf{Z}_2 \tag{3.54}$$

where $\mathbf{B} = \mathbf{F}_C^H \mathbf{Z}_1'$ is a real valued beam-space manifold of size $C \times L$. The mth element ($0 \leq m \leq C - 1$) and the lth element ($1 \leq l \leq L$) of the DFT beam-space manifold $\mathbf{B}$ is [29]

$$b_m(\nu_l) = \frac{\sin[\frac{C}{2}(\nu_l - m\frac{2\pi}{C})]}{\sin[\frac{1}{2}(\nu_l - m\frac{2\pi}{C})]}. \tag{3.55}$$

By comparing two successive beams $b_m(\nu_l)$ and $b_{m+1}(\nu_l) = \frac{\sin[\frac{C}{2}(\nu_l-(m+1)\frac{2\pi}{C})]}{\sin[\frac{1}{2}(\nu_l-(m+1)\frac{2\pi}{C})]}$, it can be shown that they are related as

$$\sin[\frac{1}{2}(\nu_l - m\frac{2\pi}{C})]b_m(\nu_l) + \sin[\frac{1}{2}(\nu_l - (m+1)\frac{2\pi}{C})]b_{m+1}(\nu_l) = 0. \tag{3.56}$$

After some trigonometric manipulations, (3.56) can be written as

$$\begin{aligned}\tan(\nu_l/2)\{\cos(m\frac{\pi}{C})b_m(\nu_l) + \cos((m+1)\frac{\pi}{C})b_{m+1}(\nu_l)\} = \\ \sin(m\frac{\pi}{C})b_m(\nu_l) + \sin((m+1)\frac{\pi}{C})b_{m+1}(\nu_l).\end{aligned} \tag{3.57}$$

It is worth mentioning that (3.57) can be found from any two beams, not necessary to be successive, where the ith invariance could be $1 \leq i \leq C-1$. Consequently, two selection matrices Γ_1 and Γ_2 of size $(C \times C)$ should be calculated to get the matrix pencil equation. For the ith invariance ($1 \leq i \leq C - 1$), the ath row of $\Gamma_{1,i}$ ($1 \leq a \leq C$) has all its elements equal to zero except the ath and the bth elements that are given by [29], [92]

$$x_a = \cos((a-1)\pi/C) \tag{3.58a}$$

$$x_b = \begin{cases} (-1)^{i+1}\cos((a+i-1)\pi/C) \, ; 1 \leq a \leq C-i, \, b = a+i \\ (-1)^{C+i}\cos((a+i-1)\pi/C) \, ; a > C-i, \, b = a+i-C \end{cases} \tag{3.58b}$$

The ath row of $\Gamma_{2,i}$ is expressed in the same way as for $\Gamma_{1,i}$ by replacing cosine functions by sine functions. For the SBMP algorithm, a single invariance is used to generate Γ_1 and Γ_2. For the MBMP algorithm, all or a number of invariances are used to generate Γ_1 and Γ_2, where Γ_1 and Γ_2 are formed by vertically staking $\Gamma_{1,i}$ and $\Gamma_{2,i}$ for $1 \leq i \leq C - 1$,

Chapter 3. System Model and Time Delay Estimation using 1-D Matrix Pencil Algorithms

respectively. By combining all C equations of (3.57) in vector form $0 \leq m \leq C - 1$, it can be written as

$$\tan(\nu_l/2)\Gamma_1 \mathbf{b}_C(\nu_l) = \Gamma_2 \mathbf{b}_C(\nu_l) \tag{3.59}$$

where $\mathbf{b}_C(\nu_l) = \{b_0(\nu_l), b_1(\nu_l), \cdots, b_{C-1}(\nu_l)\}^T$. Now, if the number of effective paths is L, the beam-space matrix of channel profile time delays is

$$\mathbf{B} = [\mathbf{b}_C(\nu_1), \mathbf{b}_C(\nu_2), \cdots, \mathbf{b}_C(\nu_L)]_{C \times L}. \tag{3.60}$$

Hence, the beam-space manifold relation of (3.59) including all time delays of the effective paths is

$$\Gamma_1 \mathbf{B} \Omega = \Gamma_2 \mathbf{B} \tag{3.61}$$

where $\Omega = \text{diag}\{\tan(\frac{\nu_1}{2}), \tan(\frac{\nu_2}{2}), \cdots, \tan(\frac{\nu_L}{2})\}$. From the definitions of $\mathbf{A}$ and $\mathbf{Z}_2$, it is clear that they are full rank matrices. Therefore, $\mathbf{B}$ and $\mathbf{Y}_F$ matrices share the same column space [29].

The real data matrix can be obtained by applying the conjugate centro-symmetrized version of the DFT matrix as

$$\mathbf{Y}_F = \mathbf{F}^H \mathbf{Y} \Rightarrow \mathbf{Y}_{\text{Re}} = [\text{Re}(\mathbf{Y}_F), \text{Im}(\mathbf{Y}_F)]. \tag{3.62}$$

For noisy data, the largest $\hat{L}$ singular left vectors of the real valued matrix $\mathbf{Y}_{\text{Re}}$ will span the column space of $\mathbf{B}$, where $\hat{L}$ is estimated by using the modified MDL. Hence, by defining $\mathbf{U}_s$ as the left singular vectors, we have

$$\mathbf{U}_s = \mathbf{BT} \tag{3.63}$$

where $\mathbf{T}$ is a nonsingular $\hat{L} \times \hat{L}$ matrix. From (3.61) and (3.63), the required matrix pencil equation of BMP is

$$\Gamma_1 \mathbf{U}_s \Psi_\nu = \Gamma_2 \mathbf{U}_s \tag{3.64}$$

where $\Psi_\nu = \mathbf{T}^{-1} \Omega \mathbf{T}$.

Summary: The summary of the enhanced 1-D BMP for channel profile parameters estimation is described in the following:

Step1: Find the Hankel matrix $\mathbf{Y}$ from the CFR as in (3.15).
Step2: Apply the conjugate centro-symmetrized version of the DFT matrix $\mathbf{Y}_F = \mathbf{F}^H \mathbf{Y}$ as in (3.62) to get the real data matrix $\mathbf{Y}_{\text{Re}} = [\text{Re}(\mathbf{Y}_F), \text{Im}(\mathbf{Y}_F)]$.
Step3: Perform the SVD on $\mathbf{Y}_{\text{Re}}$ as in (3.29) to get the signal subspace, and estimate the number of effective paths $\hat{L}$ using the modified MDL criterion as in (3.30).
Step4: Determine $\mathbf{U}_s$ as the left singular vectors of $\mathbf{U}$, which span the signal subspace and correspond to the largest $\hat{L}$ singular values of $\mathbf{Y}_{\text{Re}}$.
Step5: Calculate the selection matrices Γ_1 and Γ_2 as in (3.58) just once and store them as constant matrices.
Step6: Compute the generalized eigenvalue problem. The matrix pencil equation is

$$\Gamma_1 \mathbf{U}_s \Psi_\nu = \Gamma_2 \mathbf{U}_s \Rightarrow \Psi_\nu = (\Gamma_1 \mathbf{U}_s)^\dagger \Gamma_2 \mathbf{U}_s = (\mathbf{U}_s^H \mathbf{G}_1 \mathbf{U}_s)^{-1} \mathbf{U}_s^H \mathbf{G}_2 \mathbf{U}_s, \tag{3.65}$$

and the pencil pair is $(\Gamma_2 \mathbf{U}_s, \Gamma_1 \mathbf{U}_s)$, where $\mathbf{G}_1 = \Gamma_1^H \Gamma_1$ and $\mathbf{G}_2 = \Gamma_1^H \Gamma_2$. $\mathbf{G}_1$ and $\mathbf{G}_2$ are $C \times C$ matrices that can be calculated just once and stored as constant matrices. The eigenvalues of $\Psi_\nu = \mathbf{W} \mathbf{Z}_d \mathbf{W}^{-1}$ are computed to get $\mathbf{Z}_d = \text{diag}\{\Lambda_l = \tan(\nu_l/2);$

$l = 1, \ldots, \hat{L}\}$, where $\nu_l = -2\pi\Delta f \hat{\tau}_l$. The time delays can then be calculated as, $\hat{\tau}_l = -1/\pi\Delta f \times \tan^{-1}(\Lambda_l)$, where $l = 1, \ldots, \hat{L}$. From (3.65), the computational complexity of SBMP and MBMP algorithms are the same in real time. Such as in 1-D MP, the channel gains $\mathbf{A}_0 = [\alpha_1 e^{-j\omega_c \tau_1}, \ldots, \alpha_{\hat{L}} e^{-j\omega_c \tau_{\hat{L}}}]^T$ can be obtained by solving a least square problem as in (3.35).

3.8 Diversity Techniques using 1-D Matrix Pencil Algorithms

If a number of LTFs are available in the OFDM frame or if a number of antennas are used per each BS, a number of CFRs can be obtained. The Hankel matrix similar to (3.15) is constructed for each snapshot separately. Then, for spatial or temporal diversity, we propose the following enhanced matrix

$$\mathbf{Y}_E = \begin{bmatrix} \mathbf{Y}_0 & \mathbf{Y}_1 & \cdots & \mathbf{Y}_{q-1} \end{bmatrix}_{N-P \times q(P+1)} \quad (3.66)$$

where q is the number of estimated CFRs per each BS either by using a number of antennas for a space diversity or by using a number of LTSs per OFDM frame for a time diversity. From (3.66), the required Hankel matrices are generated first separately and then combined horizontally. The horizontal stacking has been preferred to reduce the complexity of the SVD. The size of the multiple snapshot matrix $\mathbf{Y}_E$ is $(N-P) \times q(P+1)$; the number of columns is multiplied by q. Similar to the single snapshot MP, the matrices $\mathbf{Y}_{E1}$ and $\mathbf{Y}_{E2}$ can be formed by deleting the last and the first row of (3.66), respectively. An extended matrix $\mathbf{Y}_{ex,q}$ can also be defined from (3.66) as

$$\mathbf{Y}_{ex,q} = [\mathbf{Y}_E : \prod_{N-P} \mathbf{Y}_E^* \prod_{q(P+1)}]_{(N-P) \times 2q(P+1)}. \quad (3.67)$$

The multiple snapshot forward-backward matrix $\mathbf{Y}_{ex,q}$ can also be defined by finding the extended matrix for each snapshot as in (3.27) and then combined horizontally as in (3.66). The same procedure of various 1-D MP algorithms is then applied. However, the computational complexity in SVD calculation increases if the number of snapshots in $\mathbf{Y}_E$ increases.

For large bandwidths of 802.11ac, the number of required operations is high, therefore, large bandwidths can be treated as small bandwidths by partitioning as a kind of frequency diversity. The Hankel matrix or the extended matrix is generated for each partition such as the previous, and then they are combined horizontally. For example, in case of 160 MHz BW, it can be processed as two 80 MHz bandwidths. Two Hankel or extended matrices are generated, one for the negative 80 MHz, $\mathbf{Y}_{neg}$, and one for the positive 80 MHz, $\mathbf{Y}_{pos}$. By using the multiple snapshot principle in (3.66), they are combined as

$$\mathbf{Y}_E = [\mathbf{Y}_{neg}, \mathbf{Y}_{pos}]. \quad (3.68)$$

Using the principle of frequency diversity presented in (3.68) is a robust idea, where the whole information (the measured CFR samples) can be used with a huge reduction in the complexity as we will see in the next section and a negligible reduction in the performance as we will see in the measurement chapter. The principle of temporal and spectral diversity has been presented in [54] and [55] and the principle of spatial diversity has been presented in [56].

Fig. 3.6 shows a block diagram of the proposed TDOA estimation algorithms.

3.9 Computational Complexity

In all 1-D MP algorithms, the most computationally intensive step is to estimate the signal subspace using the SVD. In 1-D UMP algorithms, the Hankel matrix $\mathbf{Y}$ of size $C \times (P+1)$ or the extended matrix $\mathbf{Y}_{ex}$ of size $C \times 2(P+1)$ is transformed by UMT into a real matrix of the same size with negligible computational effort. In 1-D BMP algorithms, the $\mathbf{F}_C^H$ matrix is multiplied by the Hankel matrix $\mathbf{Y}$ to get a real matrix of size $C \times 2(P+1)$. However, the utility of DFT beam-space MP over unitary MP is associated with DOA scenarios, where one employs a subset of the rows of $\mathbf{F}_C^H$ to transform from element space to beam-space, when some priori information is available. From (3.66) and (3.67), the number of columns of the above matrices increases by a factor of q if a number of q snapshots are used.

The computational complexity of these algorithms for estimating the singular values and left singular vectors is shown in Table 3.4 [83]. From Table 3.4, the SVD computational complexity of UMP and BMP algorithms decreases by a factor of 4. The pencil parameter was selected to be $2N/3$, where $N = N_p + N_{dc}$, to reduce the complexity and to stay in the appropriate range. Table 3.5 shows the value of N and P of each BW, where N at 20 MHz BW is equal to 53 and 57 in case of using 802.11a and 802.11n, respectively. To reduce the complexity of 160 MHz BW, it can be treated as two snapshots of 80 MHz BW, 80+80 MHz, as a kind of frequency diversity. Fig. 3.7 shows a comparison of the computational complexity of the initial transformation and SVD of these MP algorithms regarding 802.11a, 802.11n, and 802.11ac BWs based on Table 3.4 and Table 3.5.

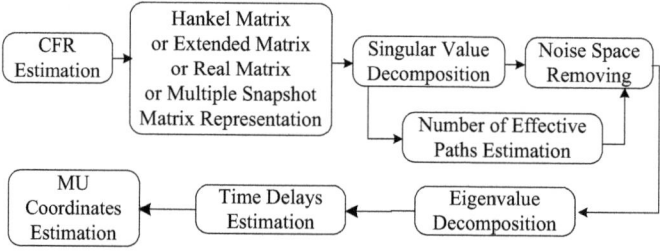

Figure 3.6: Operational flow of the proposed TDOA estimation using 1-D MP algorithms.

3.9. Computational Complexity

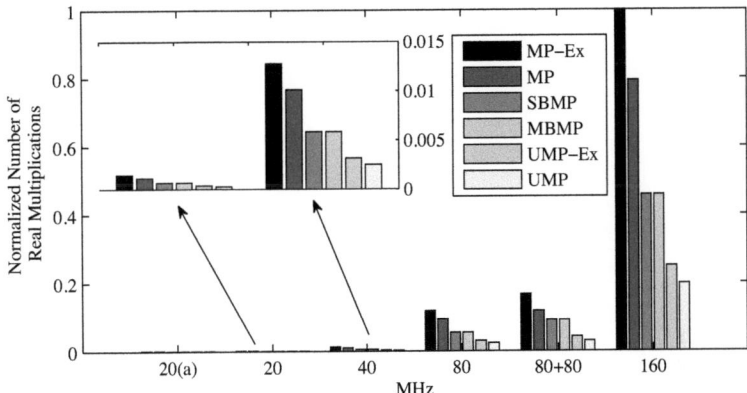

Figure 3.7: Comparison of computational complexity of the initial transformation and SVD of various 1-D MP algorithms.

Table 3.4: Computational Complexity of the Initial Transformation and SVD.

Description	Data matrix transformation	SVD
MP	$\mathbf{Y}$ no	$17C^3/3 + C^2P'q$ Complex multiplications
MP-Ex	$\mathbf{Y}_{ex}$ no	$17C^3/3 + 2C^2P'q$ Complex multiplications
UMP	$\mathbf{Y} \to \mathbf{Y}_{Re}$ Negligible Computations	$17C^3/3 + C^2P'q$ Real Multiplications
UMP-Ex	$\mathbf{Y}_{ex} \to \mathbf{Y}_{Re}$ Negligible Computations	$17C^3/3 + 2C^2P'q$ Real Multiplications
SBMP	$\mathbf{Y} \to \mathbf{Y}_F \to \mathbf{Y}_{Re}$ $C^2P'q$ Complex multiplications	$17C^3/3 + 2C^2P'q$ Real Multiplications
MBMP	$\mathbf{Y} \to \mathbf{Y}_F \to \mathbf{Y}_{Re}$ $C^2P'q$ Complex multiplications	$17C^3/3 + 2C^2P'q$ Real Multiplications
	$C = N - P, P' = P + 1$	

Table 3.5: CFR vector length and pencil parameter value regarding the 20 MHz BW of 802.11a and all BWs of 802.11ac.

Bandwidth (MHz)	20 (a)	20	40	80	160
FFT/IFFT order	64	64	128	256	512
N	53	57	117	245	501
P	35	38	78	163	334
$P' = P + 1$	36	39	79	164	335
$C = N - P$	17	19	39	82	167

CHAPTER 4

Joint Time Delay and DOA Estimation using 2-D Matrix Pencil Algorithms

Using multiple antennas at both the transmitter and receiver known by MIMO technology has attracted attention in modern wireless communications, because it offers a significant increases in data throughput and it improves the link reliability without additional bandwidth or increased transmit power. For example, the emerging IEEE 802.11ac provides as a maximum 8×8 MIMO antenna configuration [48]. While the main advantage of MIMO is to enhance data throughput, it can also be used to estimate the DOA and as a kind of spatial diversity for TDOA estimation besides the frequency diversity coming from the OFDM.

The problem of joint estimation of angles and relative time delays of multipath signals has been addressed for narrowband signals in [27]. The ESPRIT technique was used. It was assumed that the number of paths is small and discrete. The ESPRIT algorithm was also used in [107] to estimate the DOA for OFDM systems. The principle of 1-D MP was enhanced in [36] for estimating 2-D frequencies for narrowband signals using the URA. Recently, the 2-D UMP was presented in [37] to find the DOAs of the narrowband signals (azimuth and elevation angles), in which the complexity of the computations can be reduced by doing real valued computations using the UMT [94]. Furthermore, the 2-D BMP is presented in [38] to find the DOAs of the narrowband signals (azimuth and elevation angles), in which the complex data matrix can be transformed into a real data using the DFT matrix transformation; the dimension of the data matrix can also be reduced using selected rows of the DFT matrix.

In this chapter, the problem of jointly estimating the propagation time delays and relative DOAs of multipath signals has been investigated using a single transmitter and a number of receiving BSs for 2-D wireless indoor positioning system based on the hybrid TDOA and DOA measurements. Different realizations of the recent subspace-based algorithms, represented by 2-D MP algorithms, to estimate these parameters simultaneously are implemented based on the physical MIMO-OFDM system parameters. The key element of our work is to use the preamble of the OFDM frame to measure the channel state for additional purposes to the demodulation of the data portion. The ULA is used, which consists of M identical and omni-directional antennas. Hence, we have a number of M antennas distributed equally in the space dimension, and a number of N pilots distributed equally in the frequency dimension. In fact, using the ULAs and the multi-carrier signals for joint estimation of these parameters can resolve a larger number of paths in case two or more paths have equal time delays or DOAs. In addition, a high accuracy can be obtained without using high-order antenna arrays or a very wide bandwidth. Some results have been introduced in [57], and the others have been presented in [58]. The performance of using multiple antennas and wideband orthogonal multi-carrier signals are presented.

The rest of the chapter is organized as follows: first, the principle of the URA and

S-CFR are investigated in Sections 4.1 and 4.2, respectively. After that, the enhanced 2-D MP, 2-D UMP, and 2-D BMP algorithms are presented in Sections 4.3, 4.4, and 4.5, respectively. The derivations of matrix pencil equations of these algorithms can be obtained from the relevant text of each algorithm. We will focus here on the main steps with our modifications to estimate time delays and relative DOAs of wireless multipath channel. The diversity techniques using 2-D MP algorithms are presented in Section 4.6. The computational complexity of various 2-D MP algorithms is investigated precisely in Section 4.7, and it will be compared with that of 1-D MP algorithms in Section 4.8.

4.1 2-D Uniform Rectangular Array Problem Formulation

The 2-D MP algorithms were used in the literature to estimate the azimuth and elevation angles or two frequencies. In this work, they will be used to estimate the time delays and the relative DOAs of multipath propagation signals based on the estimated S-CFR. The symmetry of the S-CFR and the URA represents our key to propose that. To show that let us consider a URA of $M \times N$ sensors lying in the XY plane as shown in Fig. 4.1. The distances between array elements are Δx and Δy along the x and y directions, respectively. If L narrowband signals with wavelength λ arrive at the input of this array, the noiseless data $r(m,n)$ measured at the feeding point of the omni-directional antenna located at (m,n) can be expressed as [37], [38]

$$r(m,n) = \sum_{l=1}^{L} \alpha_l x_l^m y_l^n \tag{4.1}$$

where α_l represents the complex gain of the lth signal or the lth path, and the poles x_l and y_l are given by

$$x_l = e^{-j2\pi\Delta x \cos\phi_l \sin\theta_l/\lambda} \tag{4.2}$$

$$y_l = e^{-j2\pi\Delta y \sin\phi_l \sin\theta_l/\lambda} \tag{4.3}$$

where ϕ_l and θ_l are the azimuth and the elevation angles, respectively, as shown in Fig. 4.1. The system poles which should be estimated are

$$x_l = e^{-j\mu_l} \quad y_l = e^{-j\nu_l} \tag{4.4}$$

where $\mu_l = 2\pi\Delta x \cos\phi_l \sin\theta_l/\lambda$ and $\nu_l = 2\pi\Delta y \sin\phi_l \sin\theta_l/\lambda$.

4.2 MIMO-OFDM System Model

We consider the ULA, which contains M identical and omni-directional antennas as shown in Fig. 4.2. From chapter 2, the multipath CIR $h(t)$ is given in (2.1) as,

$$h(t) = \sum_{l=1}^{L} \alpha_l a(\theta_l) \delta(t - \tau_l).$$

For ULA, θ_l is defined as the angle between the direction orthogonal to the array axis and the impinging wave. The array response of the mth antenna to the lth path can be

4.2. MIMO-OFDM System Model

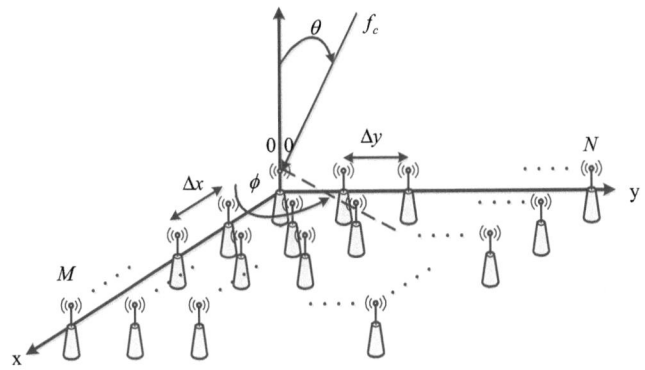

Figure 4.1: Signal modeling at the input of the URA.

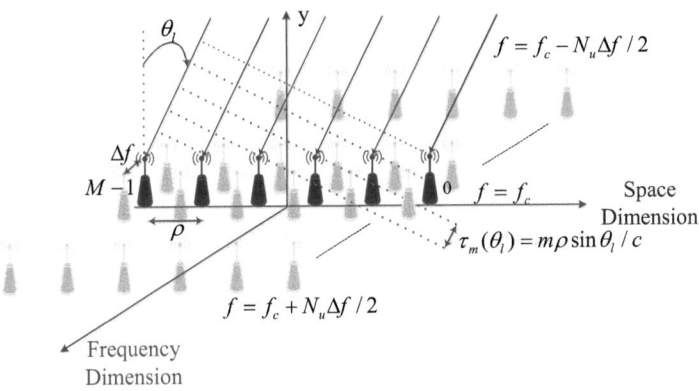

Figure 4.2: The ULA and the virtual OFDM sensors (pilots) representation. The spatial distance between array elements is ρ, and the frequency distance between OFDM sensors is Δf.

represented as in (2.2), $a_m(\theta_l) = e^{-j2\pi f \tau_m(\theta_l)}$, where $\tau_m(\theta_l) = m\rho \sin\theta_l / c$ represents the different propagation time that the plane wave impinging from direction θ_l needs to span the different distance between antenna m and the reference antenna in the antenna array as shown in Fig. 4.2. ρ is the distance between adjacent antenna elements which should satisfy $\rho \leq \lambda/2$ to avoid spatial aliasing [108], where λ is the wavelength. The vector collecting all phases of the received signals at the array elements is known as steering vector or array manyfold and is given by [65]

$$\mathbf{a}(\theta_l) = [1, e^{-j2\pi f \rho \sin(\theta_l)/c}, \ldots, e^{-j2\pi f(M-1)\rho \sin(\theta_l)/c}]. \tag{4.5}$$

From the above, the multipath propagation model represents a combination between a number of rays, each described by an arrival angle, a time delay, and a fading parameter. As it has been stated in Chapter 2, the estimation of time delay and relative angle of the first path represents our concern for positioning purposes. It represents a key element in this work to reduce the complexity and to mitigate the pairing problem in case two or more paths have equal DOAs or time delays, as it will be investigated in this chapter.

At the receiver side, once the FFT window has been adjusted, the kth subcarrier output at the reference antenna can be represented as

$$R_{0,k} = X_k \cdot \sum_{l=1}^{L} \alpha_{0,k,l} e^{-j(\omega_c + \omega_k)\tau_l} + w_{0,k} \; ; -N_u/2 \leq k \leq N_u/2 \tag{4.6}$$

where X_k is the transmitted symbol (could be a pilot, data, or a null symbol) of the kth subcarrier, $\omega_c = 2\pi f_c$ is the carrier angular frequency, $w_{0,k}$ is the AWGN at the reference antenna and the kth subcarrier, and $N_u + 1$ is the number of useful subcarriers at the central spectrum. It has been assumed that the Doppler shifts are much smaller than the subcarrier spacing. From Fig. 4.2, the signals captured by the array elements differ from each other by a phase offset due to the different propagation path. Hence, the received signal by an antenna m in the antenna array is

$$R_{m,k} = X_k \cdot \sum_{l=1}^{L} \alpha_{m,k,l} e^{-j(\omega_c + \omega_k)\tau_l} e^{-j(\omega_c + \omega_k)\tau_m(\theta_l)} + w_{m,k}. \tag{4.7}$$

For synchronization and channel parameters estimation, N_p pilot subcarriers per OFDM symbol are assumed $\{S_k : k = 0, ..., N_P - 1\}$. In another way, N_p represents the number of occupied subcarriers. It should be larger than the number of array elements M and multipath signals L to generate a generation set of the signal space [107]. The OFDM symbol structure of the training sequence has N_p symbols, and N_{dc} zero values at dc in the middle of the training sequence as it has been presented in Table 3.1 [48], [54]. For simplicity, let us mention to the minimum frequency among subcarriers by f_0 and $f_k = f_0 + k\Delta f$, where $k = 0, ..., N - 1$, and $N = N_p + N_{dc}$. The least square estimate of the CFR can be obtained from (4.7) for the mth antenna as

$$H_{m,k} = \sum_{l=1}^{L} A_{m,k,l} \cdot e^{-j2\pi k \Delta f \tau_l} e^{-j2\pi f_k m\rho \sin\theta_l / c} + w_{m,k}/S_k \tag{4.8}$$

where $A_{m,k,l} = \alpha_{m,k,l} \cdot e^{-j2\pi f_0 \tau_l}$. To show the phase differences across the frequency dimension and the space dimension, which represent the key element to use the 2-D MP

4.2. MIMO-OFDM System Model

algorithms, let us write the noiseless S-CFR in a matrix from as

$$\mathbf{H} = \begin{bmatrix} \sum_{l=1}^{L} A_{0,0,l} & \sum_{l=1}^{L} A_{0,1,l} e^{-j2\pi \Delta f \tau_l} \\ \sum_{l=1}^{L} A_{1,0,l} e^{-j2\pi f_0 \rho \sin \theta_l / c} & \sum_{l=1}^{L} A_{1,1,l} e^{-j2\pi \Delta f \tau_l} e^{-j2\pi f_1 \rho \sin \theta_l / c} \\ \vdots & \vdots \\ \sum_{l=1}^{L} A_{M-1,0,l} e^{-j2\pi f_0 (M-1)\rho \sin \theta_l / c} & \sum_{l=1}^{L} A_{M-1,1,l} e^{-j2\pi \Delta f \tau_l} e^{-j2\pi f_1 (M-1)\rho \sin \theta_l / c} \end{bmatrix}$$

$$\begin{matrix} \cdots & \sum_{l=1}^{L} A_{0,N-1,l} e^{-j2\pi \Delta f (N-1)\tau_l} \\ \cdots & \sum_{l=1}^{L} A_{1,N-1,l} e^{-j2\pi \Delta f (N-1)\tau_l} e^{-j2\pi f_{N-1} \rho \sin \theta_l / c} \\ \ddots & \vdots \\ \cdots & \sum_{l=1}^{L} A_{M-1,N-1,l} e^{-j2\pi \Delta f (N-1)\tau_l} e^{-j2\pi f_{N-1} (M-1)\rho \sin \theta_l / c} \end{matrix} \Bigg]_{M \times N}$$

(4.9)

The noiseless $H_{m,k}$ sample in terms of f_c is

$$H_{m,k} = \sum_{l=1}^{L} \beta_{m,k,l} \cdot e^{-j2\pi k \Delta f (\tau_l + m\rho \sin \theta_l / c)} e^{-j2\pi f_c m\rho \sin \theta_l / c} \quad (4.10)$$

where $-(N-1)/2 \leq k \leq (N-1)/2$, and $\beta_{m,k,l} = \alpha_{m,k,l} \cdot e^{-j2\pi f_c \tau_l}$. From (4.10), the effect of time delay coming from the antenna array is negligible compared with τ_l and it can be reduced by set the coordinates of each BS to its antenna array center, which will be considered in Section 4.2.1. Therefore, the multipath channel poles (to be estimated) can be approximated as

$$x_l = e^{-j\mu_l} \quad z_l = e^{-j\upsilon_l} \quad (4.11)$$

where

$$\mu_l = 2\pi f_c \rho \sin \theta_l / c \quad \upsilon_l = 2\pi \Delta f \tau_l. \quad (4.12)$$

The objective is to find the (x_l, z_l) pairs presented in (4.11), which correspond to the angle of arrivals and the relative time delays. It should be noted that from (4.10), for time delay estimation only, the estimated CFRs across a number of antennas can be used as a number of spatial snapshots as in Section 3.8. It represents a spatial diversity to solve the loss of rank against multipath fading. Furthermore, for DOA estimation only, the estimated CFRs across a number of subcarriers can be used as a number of spectral snapshots to solve the loss of rank against frequency-selective fading. In both cases, the principle of 1-D MP algorithms presented in the previous chapter can be used.

4.2.1 Uniform Linear Antenna Array Design

To design a ULA, let us discuss the following. The OFDM spectrum covers the frequency range of $f_c - \frac{N_u}{2}\Delta f \leq f_c \leq f_c + \frac{N_u}{2}\Delta f$. In this work, the carrier frequency f_c is used to design the antenna array. The distance between array elements is $\rho = \lambda/2 = c/2f_c$. Let us select the carrier frequency f_c to be 5.25 GHz from the frequency spectrum of IEEE

802.11 standards. The wave length will be 5.71 cm, consequence, the distance between array elements is 2.857 cm. Fig. 4.3 shows the designed ULA using 8 omni-directional antennas. It has been assumed that the time delay coming from the antenna array elements is negligible compared with the time delay coming from the wireless channel. Therefore, to show the validity of our assumption, let us discuss the following example. Assume the distance between the transmitter and the reference antenna in the antenna array is 6 meters. The time delay of the shortest path to the reference antenna in the antenna array (m = 0 as shown in Fig. 4.2) is $\tau_0 = 20ns$ while the time delay between the transmitter antenna and antenna number $m = M - 1$ in the antenna array is $\tau_{M-1} = \tau_0 + (M-1)\rho\sin(\theta_1)/c$. The largest difference in the time delay between those two antennas $m = 0$ and $m = M - 1$ occurs when the DOA of the shortest path θ_1 is $90°$. The time delay of antenna number $M - 1$ is then $\tau_{M-1} = \tau_0 + (M-1)\rho/c$, hence, in case of using 8 antennas, it will be $\tau_{M-1} = 20 + 0.667$ns. Then, the phase of the first pole is $2\pi\Delta f(20 + 0.667ns)$. Clearly, the time delay coming from the wireless channel to the reference antenna is larger than the time delay between array elements by more than ten times, which is valid as an engineering approximation. The effect of time delays coming from antenna elements could also be reduced by setting the coordinates of each BS to its antenna array center.

To show how the actual DOA and TOA values can be changed with respect to the reference coordinates of the antenna array. Fig. 4.4 shows the actual DOA and TOA if the coordinates of BS are the coordinates of the reference antenna, the antenna array center, or the last antenna in the antenna array. The distance between antenna array elements is 2.857 cm. It is clear from Fig. 4.4 that in case of using DOA in the indoor positioning system, the distance between the transmitter and the receiver is not very large to let us assume that the transmitter is in the far field and then the DOA is almost constant per each antenna. Therefore, the best option is to use the coordinates of the antenna array center of each BS.

4.3 2-D Matrix Pencil Algorithm

In the literature, the 2-D MP algorithms have been presented based on the uniform space between array elements using the URA to estimate multiple frequencies or azimuth and elevation angles. In this work, the 2-D MP algorithms are used to estimate the time delays and the relative DOAs using the OFDM and the ULA, where the first dimension is the OFDM pilots, distributed equally in the frequency dimension, and the second dimension is the array elements, distributed equally in the space dimension, which has been published in [57]. Basically, the 2-D problem of 2-D MP algorithm could be divided into two 1-D problems to estimate the poles of each dimension individually. After that the estimated poles of multipath channel should be paired to get the time delay and the relative DOA of each effective path. The single snapshot of S-CFR defined in (4.9) can be written as

$$\mathbf{H} = \begin{bmatrix} H_{0,0} & H_{0,1} & \cdots & H_{0,N-1} \\ H_{1,0} & H_{1,1} & \cdots & H_{1,N-1} \\ \vdots & \vdots & \ddots & \vdots \\ H_{M-1,0} & H_{M-1,1} & \cdots & H_{M-1,N-1} \end{bmatrix}_{M \times N}. \quad (4.13)$$

4.3. 2-D Matrix Pencil Algorithm

Figure 4.3: The designed ULA composing 8 elements.

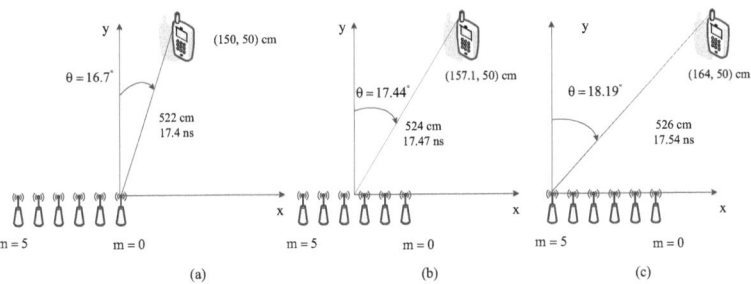

Figure 4.4: The affect of ULA coordinates selection on the values of DOA and TOA.

66 Chapter 4. Joint Time Delay and DOA Estimation using 2-D Matrix Pencil Algorithms

The above equation can be decomposed as

$$\mathbf{H} = \mathbf{A}_s \mathbf{X} \mathbf{A} \mathbf{Z} \mathbf{A}_f \tag{4.14}$$

where $\mathbf{A}_s$, $\mathbf{A}$, and $\mathbf{A}_f$ are diagonal matrices with sizes of $M \times M$, $L \times L$, and $N \times N$, respectively. They represent the complex channel gains coming from space diversity, wireless channel, and frequency diversity, respectively. To demonstrate the principle of 2-D MP, let us include the effect of frequency and space diversity channel gains of lth path to the wireless channel gains for mathematical simplification only. Hence, the S-CFR in (4.14) becomes

$$\mathbf{H} = \mathbf{X} \mathbf{A} \mathbf{Z} \tag{4.15}$$

where

$$\mathbf{X} = \begin{bmatrix} 1 & 1 & \cdots & 1 \\ x_1 & x_2 & \cdots & x_L \\ \vdots & \vdots & \ddots & \vdots \\ x_1^{M-1} & x_2^{M-1} & \cdots & x_L^{M-1} \end{bmatrix}_{M \times L}, \tag{4.16}$$

$$\mathbf{Z} = \begin{bmatrix} 1 & z_1 & \cdots & z_1^{N-1} \\ 1 & z_2 & \cdots & z_2^{N-1} \\ \vdots & \vdots & \ddots & \vdots \\ 1 & z_L & \cdots & z_L^{N-1} \end{bmatrix}_{L \times N}. \tag{4.17}$$

From (4.15), the rank of $\mathbf{H}$ doesn't exceed than L; i.e., rank($\mathbf{H}$) $\leq L$. Therefore, z_l and x_l cannot be both obtained from the principle left or right singular vectors of $\mathbf{H}$. Furthermore, the principle singular vectors of $\mathbf{H}$ do not contain sufficient information to perform the pairing between x_l and z_l [36]. To enhance the rank condition of a matrix, or in another way, to restore the dimensionality of the signal subspace, a partition and stacking process should be used besides the frequency-space smoothing. Doing that represents a core of matrix pencil algorithms rather than using a number of temporal snapshots. By using the principle of 1-D MP algorithm presented in the previous chapter, the Hankel matrix $\mathbf{Y}_m$ can be created for each antenna by windowing each row in (4.13) individually

$$\mathbf{Y}_m = \begin{bmatrix} H_{m,0} & H_{m,1} & \cdots & H_{m,N-P} \\ H_{m,1} & H_{m,2} & \cdots & H_{m,N-P+1} \\ \vdots & \vdots & \ddots & \vdots \\ H_{m,P-1} & H_{m,P} & \cdots & H_{m,N-1} \end{bmatrix}_{P \times (N-P+1)} \tag{4.18}$$

where P is the pencil parameter used to obtain the Hankel matrix. For 2-D MP algorithm, an enhanced matrix can be written in a Hankel block matrix:

$$\mathbf{Y}_e = \begin{bmatrix} \mathbf{Y}_0 & \mathbf{Y}_1 & \cdots & \mathbf{Y}_{M-K} \\ \mathbf{Y}_1 & \mathbf{Y}_2 & \cdots & \mathbf{Y}_{M-K+1} \\ \vdots & \vdots & \ddots & \vdots \\ \mathbf{Y}_{K-1} & \mathbf{Y}_K & \cdots & \mathbf{Y}_{M-1} \end{bmatrix}_{KP \times (M-K+1)(N-P+1)} \tag{4.19}$$

where K is the pencil parameter used to obtain the Hankel block matrix. P and K are like two tuning parameters, which can be adjusted to increase the estimation accuracy.

4.3. 2-D Matrix Pencil Algorithm

To select the pencil parameters P and K, some necessary conditions investigated in [36] should be satisfied as

$$\begin{align} (K-1)P &\geq L \\ K(P-1) &\geq L \\ (M-K+1)(N-P+1) &\geq L \end{align} \tag{4.20}$$

which can be understood easily from the principle of the SVD (split the received signal space into signal subspace and noise subspace) and the principle of matrix pencil equation generation (delete the first rows and the last rows), as it will be shown in this chapter. In general, the sufficient conditions to select those pencil parameters are

$$\begin{align} M-L+1 &> K \geq L+1 \\ N-L+1 &> P \geq L+1. \end{align} \tag{4.21}$$

The matrix $\mathbf{Y}_m$ of (4.18) can be written as [34]

$$\mathbf{Y}_m = \mathbf{Z}_L \mathbf{A} \mathbf{X}_d^m \mathbf{Z}_R \tag{4.22}$$

where

$$\mathbf{Z}_L = \begin{bmatrix} 1 & 1 & \cdots & 1 \\ z_1 & z_2 & \cdots & z_L \\ \vdots & \vdots & \ddots & \vdots \\ z_1^{P-1} & z_2^{P-1} & \cdots & z_L^{P-1} \end{bmatrix}_{P \times L}, \tag{4.23}$$

$$\mathbf{X}_d = \mathrm{diag}\,[x_1, x_2, \ldots, x_L], \tag{4.24}$$

$$\mathbf{Z}_R = \begin{bmatrix} 1 & z_1 & \cdots & z_1^{N-P} \\ 1 & z_2 & \cdots & z_2^{N-P} \\ \vdots & \vdots & \ddots & \vdots \\ 1 & z_L & \cdots & z_L^{N-P} \end{bmatrix}_{L \times (N-P+1)}. \tag{4.25}$$

By using (4.22) in (4.19), the enhanced matrix $\mathbf{Y}_e$ becomes

$$\mathbf{Y}_e = \mathbf{E}_L \mathbf{A} \mathbf{E}_R \tag{4.26}$$

where

$$\mathbf{E}_L = \begin{bmatrix} \mathbf{Z}_L \\ \mathbf{Z}_L \mathbf{X}_d \\ \vdots \\ \mathbf{Z}_L \mathbf{X}_d^{K-1} \end{bmatrix}, \tag{4.27}$$

$$\mathbf{E}_R = \begin{bmatrix} \mathbf{Z}_R & \mathbf{X}_d \mathbf{Z}_R & \cdots & \mathbf{X}_d^{M-K} \mathbf{Z}_R \end{bmatrix}. \tag{4.28}$$

The rows of $\mathbf{E}_L$ can be shuffled by left multiplying using the shuffling matrix $\mathbf{P}$. The shuffling matrix $\mathbf{P}$ is defined in [36] as

$$\begin{align} \mathbf{P} = [&\mathbf{s}(1), \mathbf{s}(1+P), \cdots, \mathbf{s}(1+(K-1)P), \mathbf{s}(2), \mathbf{s}(2+P), \cdots \\ &\mathbf{s}(2+(K-1)P), \cdots, \mathbf{s}(P), \mathbf{s}(P+P), \cdots, \mathbf{s}(P+(K-1)P)]^T \end{align} \tag{4.29}$$

where $s(i)$ is a column vector of size KP with one at the ith position and zero everywhere else. The shuffled matrix $\mathbf{E}_{LP}$ will be [36]

$$\mathbf{E}_{LP} = \mathbf{P}\mathbf{E}_L = \begin{bmatrix} \mathbf{X}_L \\ \mathbf{X}_L \mathbf{Z}_d \\ \vdots \\ \mathbf{X}_L \mathbf{Z}_d^{K-1} \end{bmatrix} \quad (4.30)$$

where $\mathbf{Z}_d$ has been defined in (3.22), and

$$\mathbf{X}_L = \begin{bmatrix} 1 & 1 & \cdots & 1 \\ x_1 & x_2 & \cdots & x_L \\ \vdots & \vdots & \ddots & \vdots \\ x_1^{K-1} & x_2^{K-1} & \cdots & x_L^{K-1} \end{bmatrix}_{K \times L}. \quad (4.31)$$

Summary: The enhanced 2-D MP algorithm for joint time delays and DOAs estimation can be summarized in the following:

Step1: From the estimated CFR of antenna m, find the Hankel matrix $\mathbf{Y}_m$ as in (4.18). Then, the enhanced matrix $\mathbf{Y}_e$ is defined in the Hankel block matrix as in (4.19). To increase the number of snapshots inherently, an extended matrix $\mathbf{Y}_{ex}$ can also be defined in forward and backward fashion:

$$\mathbf{Y}_{ex} = [\mathbf{Y}_e : \underset{KP}{\Pi} \mathbf{Y}_e^* \underset{(M-K+1)(N-P+1)}{\Pi}] \quad (4.32)$$

where Π is the exchange matrix that reverses the ordering of the rows. It provides also some protection against loss of rank in case of equal time delays as it has been investigated in [31] using ESPRIT algorithm for narrowband signals. It should be noted that $\mathbf{Y}_e$ has a special structure due to the centro-symmetry of the ULA-OFDM such as in the URA, therefore, its column space remains unchanged during the above transformation [109]. Based on theorem 2 in appendix A, $\mathbf{Y}_{ex}$ of (4.32) is a centro-Hermitian matrix.

Step2: The SVD is used for noisy data to reduce part of the noise effect. The matrix $\mathbf{Y}_e$ of (4.19) can then be decomposed as in (3.29), $\mathbf{Y}_e = \mathbf{U}\mathbf{\Sigma}\mathbf{V}^H$, where $\mathbf{U}$ and $\mathbf{V}$ are unitary matrices, and $\mathbf{\Sigma}$ is a diagonal matrix with the SVs of $\mathbf{Y}_e$ as, $\eta_1 \geq \eta_2 \geq \cdots \geq \eta_B$, where $B = \min(KP, (M-K+1)(N-P+1))$, which is the smaller dimension of $\mathbf{Y}_e$. The matrix $\mathbf{\Sigma}$ can be split into two submatrices $\mathbf{\Sigma}_s$ and $\mathbf{\Sigma}_n$. $\mathbf{\Sigma}_s$ is $L \times L$ diagonal matrix with the L largest SVs which characterize the signal subspace as, $\eta_k = \eta_k^s + \sigma_k^2$, where $k = 1, \ldots, L$. $\mathbf{\Sigma}_n$ is $(B-L)(B-L)$ diagonal matrix with the $(B-L)$ smallest SVs which characterize the additive noise as $\eta_k = \sigma_k^2$, where $k = L+1, \ldots, B$. If the forward-backward matrix $\mathbf{Y}_{ex}$ of (4.32) is used in (3.29), $B = \min(KP, 2(M-K+1)(N-P+1))$; it will be called the 2-D MP-Ex algorithm.

Step3: Based on the ITC, the modified MDL criterion is used to estimate the signal subspace dimension $\hat{L}$ by eliminating the noise components as it has been presented in (3.30).

Step4: The SVD of $\mathbf{Y}_e$ in (4.19), or $\mathbf{Y}_{ex}$ in (4.32), can then be decomposed as in (3.31), $\mathbf{Y} = \mathbf{Y}_s + \mathbf{Y}_n = \mathbf{U}_s \mathbf{\Sigma}_s \mathbf{V}_s^H + \mathbf{U}_n \mathbf{\Sigma}_n \mathbf{V}_n^H$, where $\mathbf{U}_s$ and $\mathbf{V}_s$ are the submatrices of

4.3. 2-D Matrix Pencil Algorithm

U and V corresponding to the SVs of $\mathbf{\Sigma}_s$ and span the signal subspace, and $\mathbf{U}_n$ and $\mathbf{V}_n$ are corresponding to the SVs of $\mathbf{\Sigma}_n$ and span the noise subspace. Such as in 1-D MP, if the number of columns in $\mathbf{U}_n$ increases (the dimension of the noise subspace), more noise components can be absorbed into the noise subspace, where a larger noise subspace implies high-estimation accuracy [36]. Therefore, the selection parameters K and P should be large enough to include the signal subspace and the noise subspace dimensions. However, KP should be smaller than $(M - K + 1)(N - P + 1)$ to reduce the overall computational complexity, as it will be explained in Section 4.7.

Step5: Extracting x_l: If the condition of (4.20) is satisfied, it has been approved in [36] that range$(\mathbf{Y}_e)$ = range$(\mathbf{U}_s)$ = range$(\mathbf{E}_L)$, where both $\mathbf{U}_s$ and $\mathbf{E}_L$ have $\hat{L}$ independent columns, and hence

$$\mathbf{U}_s = \mathbf{E}_L \mathbf{T} \tag{4.33}$$

where $\mathbf{T}$ is a unique $\hat{L} \times \hat{L}$ nonsingular matrix. The principle of matrix pencil is to construct two matrices in such a way the desired poles are the generalized eigenvalues of the corresponding matrix pencil. Therefore, the matrix pencil equation can be written along the space dimension:

$$\mathbf{U}_{s2} - \eta \mathbf{U}_{s1} = \mathbf{E}_1 \mathbf{X}_d \mathbf{T} - \eta \mathbf{E}_1 \mathbf{T} = \mathbf{E}_1(\mathbf{X}_d - \eta \mathbf{I})\mathbf{T} \tag{4.34}$$

where $\mathbf{U}_{s1}$ and $\mathbf{U}_{s2}$ are obtained from $\mathbf{U}_s$ by deleting the last and the first P rows, respectively, and $\mathbf{E}_1$ is obtained from $\mathbf{E}_L$ by deleting also the last P rows. $\mathbf{I}$ is an identity matrix with a proper dimension. From (4.34), the set of $\hat{L}$ spatial poles $\{x_1, x_2, \ldots, x_{\hat{L}}\}$ of the diagonal matrix $\mathbf{X}_d$ are the rank reducing numbers of the matrix pencil $\mathbf{U}_{s2} - \eta \mathbf{U}_{s1}$. In another way, the rank of the matrix pencil decreases by one if and only if $x_l = \eta$. Therefore, the desired set of $\hat{L}$ spatial poles $\{x_l\}$ are the eigenvalues of pencil pair $(\mathbf{U}_{s2}, \mathbf{U}_{s1})$. They can be calculated as the eigenvalues of $\mathbf{\Psi}_\mu = \mathbf{U}_{s1}^\dagger \mathbf{U}_{s2}$, which represents a generalized eigenvalue problem of dimension $\hat{L} \times \hat{L}$. The lth DOA can then be calculated from $x_l = e^{-j2\pi\rho \sin\hat{\theta}_l / \lambda}$, where $l = 1, \ldots, \hat{L}$, as

$$\hat{\theta}_l = \sin^{-1}(\arg(x_l^*) . \lambda / 2\pi\rho) \tag{4.35}$$

Step6: Extracting z_l: In order to estimate $\{z_l; l = 1, \ldots, \hat{L}\}$, the structure of $\mathbf{E}_{LP} = \mathbf{P}\mathbf{E}_L$ in (4.30) is used. Hence, the shuffling matrix defined in (4.29) is used to introduce the permutation in (4.33) as

$$\mathbf{U}_{sp} = \mathbf{P}\mathbf{U}_s \Rightarrow \mathbf{U}_{sp} = \mathbf{P}\mathbf{E}_L \mathbf{T} = \mathbf{E}_{LP}\mathbf{T}. \tag{4.36}$$

The matrix pencil equation can be written along the frequency dimension as

$$\mathbf{U}_{sp2} - \eta \mathbf{U}_{sp1} = \mathbf{E}_{1p} \mathbf{Z}_d \mathbf{T} - \eta \mathbf{E}_{1p} \mathbf{T} = \mathbf{E}_{1p}(\mathbf{Z}_d - \eta \mathbf{I})\mathbf{T} \tag{4.37}$$

where $\mathbf{U}_{sp1}$ and $\mathbf{U}_{sp2}$ are obtained from $\mathbf{U}_{sp}$ by deleting the last and the first K rows, respectively, and $\mathbf{E}_{1p}$ is obtained from $\mathbf{E}_{LP}$ by deleting also the last K rows. The desired set of $\hat{L}$ spectral poles $\{z_l\}$ can be obtained as the generalized eigenvalues of pencil pair

($\mathbf{U}_{sp2}, \mathbf{U}_{sp1}$). They can be calculated as the eigenvalues of $\mathbf{\Psi}_v = \mathbf{U}_{sp1}^\dagger \mathbf{U}_{sp2}$, which represents a generalized eigenvalue problem of dimension $\hat{L} \times \hat{L}$. The lth time delay can then be calculated from $z_l = e^{-j2\pi\Delta f \hat{\tau}_l}$, where $l = 1, ..., \hat{L}$, as

$$\hat{\tau}_l = \arg(z_l^*)/2\pi\Delta f. \tag{4.38}$$

The order of poles in each set, namely $\mathbf{z} = \{z_l; l = 1, ..., \hat{L}\}$ and $\mathbf{x} = \{x_l; l = 1, ..., \hat{L}\}$, is still unknown. Since, the eigenvalue decomposition (EVD) provides the values of the eigenvalues only, and it does not provide any order for the eigenvalues [36]. In addition, if the number of effective poles has been overestimated, some of the resulting eigenvalues will be repeated. Several efforts have been made to pair the unknown parameters in 2-D scenarios, which could be multiple frequencies or azimuth and elevation angles in the literature [29], [110], [111], [112], [113], [114], [115]. In the following, two algorithms will be presented. The proposed algorithm for wireless indoor positioning will be presented, which has been published in [57].

4.3.1 Correlation Maximization Pairing Method

The order of poles in the estimated sets $\mathbf{z} = \{z_l; l = 1, ..., \hat{L}\}$ and $\mathbf{x} = \{x_l; l = 1, ..., \hat{L}\}$ is still unknown. Therefore, the two sets of estimated poles should be paired to find the corresponding time delay and the relative DOA of each path. Based on the orthogonal property between the signal subspace and the noise subspace, the pairs can be correctly paired together by maximizing the criterion below [36]

$$J_s(i,j) = \sum_{l=1}^{\hat{L}} \left\| \mathbf{u}_l^H \mathbf{e}_{ij}(x_i, z_j) \right\|^2 ; i, j = 1, \cdots, \hat{L} \tag{4.39}$$

$$\mathbf{e}_{ij} = \mathbf{x}_i \otimes \mathbf{z}_j \tag{4.40}$$

where $\otimes$ is the Kronecker product defined in appendix B, $\mathbf{x}_i = [1, x_i, ..., x_i^{K-1}]^T$, $\mathbf{z}_j = [1, z_j, ..., z_j^{P-1}]^T$, and $\{\mathbf{u}_l; l = 1, ..., \hat{L}\}$ are the principal eigenvectors. From (4.39), it can be noted that, the pairing procedure does not deal with the poles $\{x_l; l = 1, ..., \hat{L}\}$ equally, i.e., x_l is considered before x_{l+1}. In our joint time delay and DOA estimation problem, we have a priori information about z_l, where the shortest path represents our concern. Therefore, we order z_l poles according to their priority (in an ascending order), and then the pairing procedure is applied. For example, we set $j = 1$ and search over $\{x_i; i = 1, ..., \hat{L}\}$ for the best matched pole to z_1. However, besides the computational complexity, this pairing method with our modification does not always provide the correct pairing results, when there are repeated poles, because it is a correlation maximization method based on the orthogonal property between noise and signal subspaces [109], [115]. In certain scenarios, when the eigen vectors of both sets are close enough, the correlation maximization grouping method may fail especially at low SNR [109], [115].

4.3.2 Proposed Pairing Method

In this work, the principle of both the wireless indoor positioning and the proposed method in [115] are used to propose the following method. A pairing method has been

4.3. 2-D Matrix Pencil Algorithm

presented in [115] for 2-D frequencies estimation, which provides more accurate pairing results compared with the others in the literature and requires less computational complexity. It can be realized in our case as, the multipath channel poles of the frequency and space dimensions have the same generalized eigenvectors when they are decomposed using the generalized EVD. The EVD problem of both sets can then be written as

$$\mathbf{U}_{s1}^{\dagger}\mathbf{U}_{s2} = \mathbf{\Psi}_{\mu} = \mathbf{W}\mathbf{X}_d\mathbf{W}^{-1} \Rightarrow \mathbf{X}_d = \mathbf{W}^{-1}\mathbf{\Psi}_{\mu}\mathbf{W}, \quad (4.41)$$

$$\mathbf{U}_{sp1}^{\dagger}\mathbf{U}_{sp2} = \mathbf{\Psi}_{v} = \mathbf{W}\mathbf{Z}_d\mathbf{W}^{-1} \Rightarrow \mathbf{Z}_d = \mathbf{W}^{-1}\mathbf{\Psi}_{v}\mathbf{W}. \quad (4.42)$$

It is clear from the above equations that the same eigenvectors are used to estimate both $\mathbf{X}_d$ and $\mathbf{Z}_d$. This property can be utilized to estimate the required parameters in a grouped form without the need to solve an eigenvalue problem for each dimension and then use computationally expensive pairing methods. Hence, the proposed method guarantees that the estimated poles are paired up correctly.

In case of wireless positioning, our concern is to estimate the time delay of the first path and the corresponding DOA. Therefore, the eigenvalue problem of $\mathbf{\Psi}_v$ is computed first to find the eigenvalues of the diagonal matrix $\mathbf{Z}_d$ and the eigenvectors $\mathbf{W}$ as in (4.42). Then, the estimated propagation time delays of the multipath channel are rearranged in an ascending order. After that, the eigenvectors of $\mathbf{W}$ are rearranged in accordance with the previous order to get $\mathbf{W}'$. The corresponding column of the shortest path is put as the first column in $\mathbf{W}'$ since our concern is the DOA of the shortest path. Finally, compute $\mathbf{X}_d$ by premultiplying $\mathbf{\Psi}_{\mu}$ with the inverse $\mathbf{W}'^{-1}$, and then postmultiplying with $\mathbf{W}'$. From the above, the proposed method has some advantages can be summarized in the following:

- A single EVD is used instead of two to estimate both sets $\mathbf{z} = \{z_l; l = 1, ..., \hat{L}\}$ and $\mathbf{x} = \{x_l; l = 1, ..., \hat{L}\}$.

- It does not need to use a pairing method to pair the estimated poles. They are extracted simultaneously, and the ordering of their eigenvalues is corresponding.

- Since the DOA of the shortest path is our concern; the problem of repeated poles has been mitigated for wireless positioning and telecommunication applications.

Such as in 1-D MP algorithms, the channel gains can be obtained by solving a least square problem. Once both sets $\mathbf{z} = \{z_l; l = 1, ..., \hat{L}\}$ and $\mathbf{x} = \{x_l; l = 1, ..., \hat{L}\}$ are estimated, the channel gains can be obtained from (4.15) as

$$\mathbf{A} = \mathbf{X}^{\dagger}\mathbf{H}\mathbf{Z}^{\dagger}. \quad (4.43)$$

From the above, the time delays of the effective paths have been estimated, which can be used in telecommunication equalizer, and the DOA of the shortest path has also been estimated, which can be used in beam forming to improve signal quality and then to increase capacity. However, if there are a number of repeated poles and we need to estimate all DOAs of all paths in the channel profile, the EVD problem should be solved a number of times as described in [115].

4.4 2-D Unitary Matrix Pencil Algorithm

The principle of real valued computations in 1-D UMP was also extended to 2-D UMP in [37] to estimate the azimuth and elevation angles of narrowband signals using the URA. In the previous chapter, the developed 1-D UMP has been applied successfully for wireless positioning using OFDM systems, which has been published in [54], [56] and [55] for different enhancements in different scenarios. In this section, the principle of the computationally efficient 2-D UMP algorithm will be developed for wireless positioning using the wideband orthogonal multi-carrier signals represented by OFDM and the ULA to estimate the time delays and the relative DOAs of a wireless multipath channel simultaneously. Results have been published in [57], as we will see in the measurement chapter. To do that, three theorems of the UMT are required, presented in appendix A, where their proofs can be found in [37].

By using the centro-symmetry of the ULA-OFDM or any complex matrix could be written in a centro-hermitian matrix form, the centro-Hermitian matrix $\mathbf{Y}_{ex}$ defined in (4.32) can be converted to a real matrix $\mathbf{Y}_{Re}$ using theorem 3 in appendix A:

$$\mathbf{Y}_{Re} = \mathbf{Q}_{K_1}^H \mathbf{Y}_{ex} \mathbf{Q}_{K_2} \qquad (4.44)$$

where $\mathbf{Q}_{K_1}$ and $\mathbf{Q}_{K_2}$ are unitary matrices defined in (3.37) and (3.38).

After that some selection matrices such as in 1-D UMP are used to write the matrix pencil equations of the space and frequency dimensions. The selection matrices $\mathbf{J}_{u1}$ and $\mathbf{J}_{u2}$ are used to select the rows of the real matrix $\mathbf{Y}_{Re}$ in order to write the matrix pencil equation along the space dimension for $\{x_l\}$ poles estimation, which are given by

$$\mathbf{J}_{u1} = [\mathbf{I}_{KP-P} \; : \; \mathbf{0}_{(KP-P) \times P}], \qquad (4.45)$$

$$\mathbf{J}_{u2} = [\mathbf{0}_{(KP-P) \times P} \; : \; \mathbf{I}_{KP-P}]. \qquad (4.46)$$

The same for the frequency dimension, the selection matrices $\mathbf{J}_{v1}$ and $\mathbf{J}_{v2}$ are used to select the rows of the real matrix $\mathbf{Y}_{Re}$ in order to write the matrix pencil equation for $\{z_l\}$ poles estimation, which are given by

$$\mathbf{J}_{v1} = [\mathbf{I}_{KP-K} \; : \; \mathbf{0}_{(KP-K) \times K}], \qquad (4.47)$$

$$\mathbf{J}_{v2} = [\mathbf{0}_{(KP-K) \times K} \; : \; \mathbf{I}_{KP-K}]. \qquad (4.48)$$

In order to estimate $\{x_l\}$ poles, the matrix pencil equation can be written as

$$\mathbf{J}_{u2} \mathbf{Y}_{ex} = x \mathbf{J}_{u1} \mathbf{Y}_{ex}. \qquad (4.49)$$

By using the UMT principle, presented in (3.42) to (3.46), it can be found that

$$\mathbf{Q}^H \mathbf{J}_{u2} \mathbf{Q} = (\mathbf{Q}^H \mathbf{J}_{u1} \mathbf{Q})^* \qquad (4.50)$$

where the unitary matrices $\mathbf{Q}$ have proper dimensions. As it is described in the previous, $\mathbf{x} = \{x_l; l = 1, ..., \hat{L}\}$ are the rank reducing numbers of the matrix pencil equation, hence, substituting (4.44) and (4.50) into (4.49) gives

$$(\mathbf{Q}^H \mathbf{J}_{u1} \mathbf{Q})^* \mathbf{Y}_{Re} = e^{-j\mu_l} \mathbf{Q}^H \mathbf{J}_{u1} \mathbf{Q} \mathbf{Y}_{Re} \qquad (4.51)$$

4.4. 2-D Unitary Matrix Pencil Algorithm

where $\mu_l = 2\pi f_c \rho \sin\theta_l/c$ defined in (4.12). By using the definition of the tangent function with some mathematical manipulations, it can be shown that

$$\tan(\mu_l/2)\mathbf{K}_{\mathrm{Re}1}\mathbf{Y}_{\mathrm{Re}} = \mathbf{K}_{\mathrm{Im}1}\mathbf{Y}_{\mathrm{Re}} \tag{4.52}$$

where

$$\mathbf{K}_{\mathrm{Re}1} = \mathrm{Re}(\mathbf{Q}_{K_1}^H \mathbf{J}_{u1} \mathbf{Q}_{K_2}) \quad \mathbf{K}_{\mathrm{Im}1} = \mathrm{Im}(\mathbf{Q}_{K_1}^H \mathbf{J}_{u1} \mathbf{Q}_{K_2}). \tag{4.53}$$

By using the same principle of 1-D MP, the matrix $\mathbf{Y}_{\mathrm{Re}}$ can be replaced by its signal subspace $\mathbf{U}_s$ in order to reduce the effect of noise. Therefore, the matrix pencil equation of the space dimension is

$$\mathbf{K}_{\mathrm{Re}1}\mathbf{U}_s\mathbf{\Psi}_\mu = \mathbf{K}_{\mathrm{Im}1}\mathbf{U}_s. \tag{4.54}$$

The generalized eigenvalues of pencil pair $(\mathbf{K}_{\mathrm{Im}1}\mathbf{U}_s, \mathbf{K}_{\mathrm{Re}1}\mathbf{U}_s)$ can be calculated as the eigenvalues of

$$\mathbf{\Psi}_\mu = [\mathbf{K}_{\mathrm{Re}1}\mathbf{U}_s]^\dagger \mathbf{K}_{\mathrm{Im}1}\mathbf{U}_s = \mathbf{W}\mathbf{X}_d\mathbf{W}^{-1} \tag{4.55}$$

to get $\mathbf{X}_d = \mathrm{diag}\{\Omega_l = \tan(\mu_l/2); l = 1, \ldots, \hat{L}\}$.

By the same way of extracting $\{x_l\}$ poles, the matrix pencil equation for $\{z_l\}$ poles estimation can be written as

$$\mathbf{J}_{v2}\mathbf{P}\mathbf{Y}_{ex} = z\mathbf{J}_{v1}\mathbf{P}\mathbf{Y}_{ex} \tag{4.56}$$

which can be written as

$$\mathbf{Q}^H\mathbf{J}_{v2}\mathbf{Q}\mathbf{Q}^H\mathbf{P}\mathbf{Q}\mathbf{Q}^H\mathbf{Y}_{ex}\mathbf{Q} = z\mathbf{Q}^H\mathbf{J}_{v1}\mathbf{Q}\mathbf{Q}^H\mathbf{P}\mathbf{Q}\mathbf{Q}^H\mathbf{Y}_{ex}\mathbf{Q}. \tag{4.57}$$

Substituting (4.44) into (4.57) gives

$$\mathbf{Q}^H\mathbf{J}_{v2}\mathbf{Q}\mathbf{P}'\mathbf{Y}_{\mathrm{Re}} = z\mathbf{Q}^H\mathbf{J}_{v1}\mathbf{Q}\mathbf{P}'\mathbf{Y}_{\mathrm{Re}} \tag{4.58}$$

where the modified shuffling matrix is

$$\mathbf{P}' = \mathbf{Q}_{K_1}^H \mathbf{P} \mathbf{Q}_{K_2} \tag{4.59}$$

due to using the UMT. It has been illustrated in [37] and [109] that $\mathbf{P}$ remains unchanged if its rows and columns are reversed. Since $\mathbf{P}$ is a real matrix, it means that

$$\mathbf{P} = \Pi\mathbf{P}\Pi = \Pi\mathbf{P}^*\Pi. \tag{4.60}$$

From (4.60), $\mathbf{P}$ is a centro-Hermitian matrix. Therefore, based on theorem 3 in appendix A, the UMT of matrix $\mathbf{P}$ in (4.59), $\mathbf{P}'$, is always a real matrix. Similarly to (4.54), the matrix pencil equation of the frequency dimension is

$$\mathbf{K}_{\mathrm{Re}2}\mathbf{U}_{sp}\mathbf{\Psi}_v = \mathbf{K}_{\mathrm{Im}2}\mathbf{U}_{sp} \tag{4.61}$$

where $\mathbf{K}_{\mathrm{Re}2}$ and $\mathbf{K}_{\mathrm{Im}2}$ are

$$\mathbf{K}_{\mathrm{Re}2} = \mathrm{Re}(\mathbf{Q}_{K_1}^H \mathbf{J}_{v1} \mathbf{Q}_{K_2}) \quad \mathbf{K}_{\mathrm{Im}2} = \mathrm{Im}(\mathbf{Q}_{K_1}^H \mathbf{J}_{v1} \mathbf{Q}_{K_2}). \tag{4.62}$$

Like in 2-D MP, $\mathbf{U}_{sp}$ is shuffled by the shuffling matrix as, $\mathbf{U}_{sp} = \mathbf{P}'\mathbf{U}_s$, where $\mathbf{P}'$ is defined in (4.59) due to using the UMT. The EVD of pencil pair $(\mathbf{K}_{\mathrm{Im}2}\mathbf{U}_{sp}, \mathbf{K}_{\mathrm{Re}2}\mathbf{U}_{sp})$ is computed as

$$\mathbf{\Psi}_v = [\mathbf{K}_{\mathrm{Re}2}\mathbf{U}_{sp}]^\dagger \mathbf{K}_{\mathrm{Im}2}\mathbf{U}_{sp} = \mathbf{W}\mathbf{Z}_d\mathbf{W}^{-1} \tag{4.63}$$

74 Chapter 4. Joint Time Delay and DOA Estimation using 2-D Matrix Pencil Algorithms

to get $\mathbf{Z}_d = \text{diag}\{\Lambda_l = \tan(v_l/2); l = 1, \ldots, \hat{L}\}$.

Summary: The enhanced 2-D UMP algorithm for time delays and relative DOAs estimation can be summarized in the following:

Step1: Such as the first step of 2-D MP algorithm, define the centro-Hermitian matrix $\mathbf{Y}_{ex}$ as in (4.32).

Step2: Calculate the real data matrix $\mathbf{Y}_{\text{Re}}$, using the centro-symmetry of the ULA-OFDM or any complex matrix could be written in a centro-hermitian matrix form as in (4.44), $\mathbf{Y}_{\text{Re}} = \mathbf{Q}_{K_1}^H \mathbf{Y}_{ex} \mathbf{Q}_{K_2}$.

Step3: Perform the SVD on $\mathbf{Y}_{\text{Re}}$ as in (3.29) to get the signal subspace, and estimate the number of effective paths $\hat{L}$ using the modified MDL criterion presented in (3.30).

Step4: Determine $\mathbf{U}_s$ as the left singular vectors of $\mathbf{U}$, which span the signal subspace and correspond to the largest $\hat{L}$ singular values of $\mathbf{Y}_{\text{Re}}$.

Next, we calculate the selection matrices $\mathbf{K}_{\text{Re}\,2}$ and $\mathbf{K}_{\text{Im}\,2}$ to extract a set of $\hat{L}$ spectral poles $\{z_l\}$ just once and store them as constant matrices according to $\mathbf{K}_{\text{Re}\,2} = \text{Re}(\mathbf{Q}_{K_1}^H \mathbf{J}_{v1} \mathbf{Q}_{K_2})$ and $\mathbf{K}_{\text{Im}\,2} = \text{Im}(\mathbf{Q}_{K_1}^H \mathbf{J}_{v1} \mathbf{Q}_{K_2})$, where $\mathbf{J}_{v1}$ is a selection matrix constructed from an identity matrix and a zero matrix as, $\mathbf{J}_{v1} = [\mathbf{I}_{KP-K} : \mathbf{0}_{(KP-K) \times K}]$. In addition, we calculate the selection matrices $\mathbf{K}_{\text{Re}\,1}$ and $\mathbf{K}_{\text{Im}\,1}$ to extract another set of $\hat{L}$ spatial poles $\{x_l\}$ just once and store them as constant matrices according to $\mathbf{K}_{\text{Re}\,1} = \text{Re}(\mathbf{Q}_{K_1}^H \mathbf{J}_{u1} \mathbf{Q}_{K_2})$ and $\mathbf{K}_{\text{Im}\,1} = \text{Im}(\mathbf{Q}_{K_1}^H \mathbf{J}_{u1} \mathbf{Q}_{K_2})$, where $\mathbf{J}_{u1}$ is a selection matrix constructed from an identity matrix and a zero matrix as, $\mathbf{J}_{u1} = [\mathbf{I}_{KP-P} : \mathbf{0}_{(KP-P) \times P}]$.

Step5: Extracting z_l: Compute the EVD of $\mathbf{\Psi}_v = [\mathbf{K}_{\text{Re}\,2}\mathbf{U}_{sp}]^\dagger \mathbf{K}_{\text{Im}\,2}\mathbf{U}_{sp}$ as, $\mathbf{\Psi}_v = \mathbf{W}\mathbf{Z}_d\mathbf{W}^{-1}$, to get $\mathbf{Z}_d = \text{diag}\{\Lambda_l = \tan(v_l/2); l = 1, \cdots, \hat{L}\}$, where $v_l = 2\pi\Delta f \hat{\tau}_l$. Like in 2-D MP, $\mathbf{U}_{sp}$ is shuffled by the shuffling matrix as, $\mathbf{U}_{sp} = \mathbf{P}'\mathbf{U}_s$, where $\mathbf{P}'$ is $\mathbf{Q}_{K_1}^H \mathbf{P} \mathbf{Q}_{K_2}$ due to using the UMT. Then, calculate the time delays according to $\hat{\tau}_l = 1/\pi\Delta f \times \tan^{-1}(\Lambda_l)$, where $l = 1, \cdots, \hat{L}$.

Step6: Extracting x_l: By using the proposed method in 2-D MP in Section 4.3.2, compute $\mathbf{X}_d$ by premultiplying $\mathbf{\Psi}_\mu = [\mathbf{K}_{\text{Re}\,1}\mathbf{U}_s]^\dagger \mathbf{K}_{\text{Im}\,1}\mathbf{U}_s$ with the inverse $\mathbf{W}'^{-1}$, and then postmultiplying with $\mathbf{W}'$ to get $\mathbf{X}_d = \text{diag}\{\Omega_l = \tan(\mu_l/2); l = 1, \cdots, \hat{L}\}$, where $\mu_l = 2\pi\rho\sin\hat{\theta}_l/\lambda$. After that, calculate the DOAs according to $\sin\hat{\theta}_l = c/\pi f_c\rho \times \tan^{-1}(\Omega_l)$, where $l = 1, \cdots, \hat{L}$.

It is worth mentioning that after the initial transformation using the UMT, the whole processing is real valued computation.

An automatic pairing method has been presented in [29] for unitary 2-D ESPRIT to estimate the azimuth and elevation angles for narrowband signals using the URA. Its principle can be enhanced to this work. The quantities $\mathbf{\Psi}_\mu$ and $\mathbf{\Psi}_v$ are real valued matrices, hence, they can be decomposed as

$$\mathbf{\Psi}_\mu + j\mathbf{\Psi}_v = \mathbf{W}\boldsymbol{\eta}_d\mathbf{W}^{-1} = \mathbf{W}(\mathbf{X}_d + j\mathbf{Z}_d)\mathbf{W}^{-1} \tag{4.64}$$

to find the complex eigenvalues $\boldsymbol{\eta}_d = \text{diag}\{\eta_l; l = 1, \cdots, \hat{L}\}$. After that the required poles of both sets can be computed as follows

$$\mu_l = 2\tan^{-1}(\text{Re}(\eta_l)); l = 1, \cdots, \hat{L}, \tag{4.65}$$

$$v_l = 2\tan^{-1}(\text{Im}(\eta_l)); \; l = 1, \cdots, \hat{L}. \tag{4.66}$$

However, there are some disadvantages of the automatic pairing method. The first problem is that in some critical cases, it does not guarantee that $\mathbf{W}$ is a real matrix. It might happen that $\mathbf{W}$ becomes complex, for example, the decomposition of $\mathbf{\Psi}_\mu = \mathbf{W}\mathbf{X}_d\mathbf{W}^{-1}$ is complex. Hence, both estimates may corrupt each other [31]. However, it can work well in normal cases, and a single complex EVD is used. In our case of wireless positioning, the URA can be used, and hence the various 2-D MP algorithms can be extended to 3-D MP algorithms to do joint estimation of time delays and both azimuth and elevation angles, which is a future work. Therefore, the second problem in case of using the automatic pairing method is that it cannot be extended to more than two dimensions. Therefore, the proposed pairing method in Section 4.3.2 is outperformed with respect to the complexity, the accuracy, and the future work. The required poles are always estimated and grouped simultaneously using the same eigenvector matrix.

4.5 2-D Beam-space Matrix Pencil Algorithm

The DFT was applied successfully in [92] for 1-D BMP algorithm, in which the complexity of the computation is reduced by using real computations. In [53], and [54], the performance of the enhanced 1-D BMP algorithm for wireless indoor positioning using OFDM signals has been investigated, as it will be seen in the measurement chapter. The same principle of real computations in 1-D BMP was also extended to 2-D BMP in [38] to estimate the azimuth and elevation angles of the narrowband signals using the URA. In this work, the principle of 2-D BMP has also been developed for wireless positioning systems to estimate time delays and relative angle of arrivals. The derivation of 2-D BMP will be omitted here; interested readers can see the derivation in the relevant text [38]. Results have been presented in [58].

The enhanced matrix $\mathbf{Y}_e$ defined in (4.19) has been decomposed in (4.26) as, $\mathbf{Y}_e = \mathbf{E}_L \mathbf{A} \mathbf{E}_R$, where $\mathbf{E}_L$ and $\mathbf{E}_R$ can also be written as [38]

$$\mathbf{E}_L = \mathbf{X}_L \odot \mathbf{Z}_L, \tag{4.67}$$

$$\mathbf{E}_R^T = \mathbf{X}_R^T \odot \mathbf{Z}_R^T, \tag{4.68}$$

where the symbol $\odot$ denotes the Khatri-Rao product defined in (B.2) of appendix B [116], and the matrices $\mathbf{Z}_L$, $\mathbf{Z}_R$, and $\mathbf{X}_L$ are given in (4.23), (4.25), and (4.31), respectively. The matrix $\mathbf{X}_R$ is given by

$$\mathbf{X}_R = \begin{bmatrix} 1 & x_1 & \cdots & x_1^{M-K} \\ 1 & x_2 & \cdots & x_2^{M-K} \\ \vdots & \vdots & \ddots & \vdots \\ 1 & x_L & \cdots & x_L^{M-K} \end{bmatrix}_{L \times (M-K+1)}. \tag{4.69}$$

Such as the principle of 1-D BMP, the matrices $\mathbf{X}_L$ and $\mathbf{Z}_L$ can also be decomposed as

$$\mathbf{X}_L = \mathbf{X}'_L \mathbf{X}_o \tag{4.70}$$

$$\mathbf{Z}_L = \mathbf{Z}'_L \mathbf{Z}_o \tag{4.71}$$

where

$$\mathbf{X}'_L = \begin{bmatrix} x_1^{-\frac{K-1}{2}} & x_2^{-\frac{K-1}{2}} & \cdots & x_L^{-\frac{K-1}{2}} \\ x_1^{-\frac{K-3}{2}} & x_2^{-\frac{K-3}{2}} & \cdots & x_L^{-\frac{K-3}{2}} \\ \vdots & \vdots & \ddots & \vdots \\ x_1^{\frac{K-3}{2}} & x_2^{\frac{K-3}{2}} & \cdots & x_L^{\frac{K-3}{2}} \\ x_1^{\frac{K-1}{2}} & x_2^{\frac{K-1}{2}} & \cdots & x_L^{\frac{K-1}{2}} \end{bmatrix}_{K \times L}, \tag{4.72}$$

$$\mathbf{Z}'_L = \begin{bmatrix} z_1^{-\frac{P-1}{2}} & z_2^{-\frac{P-1}{2}} & \cdots & z_L^{-\frac{P-1}{2}} \\ z_1^{-\frac{P-3}{2}} & z_2^{-\frac{P-3}{2}} & \cdots & z_L^{-\frac{P-3}{2}} \\ \vdots & \vdots & \ddots & \vdots \\ z_1^{\frac{P-3}{2}} & z_2^{\frac{P-3}{2}} & \cdots & z_L^{\frac{P-3}{2}} \\ z_1^{\frac{P-1}{2}} & z_2^{\frac{P-1}{2}} & \cdots & z_L^{\frac{P-1}{2}} \end{bmatrix}_{P \times L}, \tag{4.73}$$

$$\mathbf{X}_o = \text{diag}\{x_1^{\frac{K-1}{2}}, x_2^{\frac{K-1}{2}}, \ldots, x_L^{\frac{K-1}{2}}\}, \tag{4.74}$$

$$\mathbf{Z}_o = \text{diag}\{z_1^{\frac{P-1}{2}}, z_2^{\frac{P-1}{2}}, \ldots, z_L^{\frac{P-1}{2}}\}. \tag{4.75}$$

From (4.26) and (4.70) to (4.75), $\mathbf{Y}_e$ becomes

$$\mathbf{Y}_e = (\mathbf{X}'_L \odot \mathbf{Z}'_L) \mathbf{X}_o \mathbf{Z}_o \mathbf{A} \mathbf{E}_R. \tag{4.76}$$

. Assume that $\mathbf{F}_K$ and $\mathbf{F}_P$ are the DFT matrices of dimensions $K \times K$ and $P \times P$, respectively. The mth row of $\mathbf{F}_K^H$ and the nth row of $\mathbf{F}_P^H$ can be written as

$$\mathbf{f}_m^H = e^{j\frac{K-1}{2}m\frac{2\pi}{K}} \times [1, e^{-jm\frac{2\pi}{K}}, e^{-j2m\frac{2\pi}{K}}, \cdots, e^{-j(K-1)m\frac{2\pi}{K}}] \tag{4.77}$$

$$\mathbf{f}_n^H = e^{j\frac{P-1}{2}n\frac{2\pi}{P}} \times [1, e^{-jn\frac{2\pi}{P}}, e^{-j2n\frac{2\pi}{P}}, \cdots, e^{-j(P-1)n\frac{2\pi}{P}}] \tag{4.78}$$

where the row vectors $\mathbf{f}_m^H$ and $\mathbf{f}_n^H$ represent the DFT beams steered at spatial frequencies, $\mu = m \times 2\pi/K$ and $v = n \times 2\pi/P$, respectively. The 2-D DFT matrix is then given by

$$\mathbf{F}^H = \mathbf{F}_K^H \otimes \mathbf{F}_P^H. \tag{4.79}$$

If (4.76) is multiplied by $\mathbf{F}^H$ from left, it will be

$$\mathbf{Y}_F = \mathbf{F}^H \mathbf{Y}_e = (\mathbf{F}_K^H \otimes \mathbf{F}_P^H)(\mathbf{X}'_L \odot \mathbf{Z}'_L) \mathbf{X}_o \mathbf{Z}_o \mathbf{A} \mathbf{E}_R. \tag{4.80}$$

By using the property of Kronecker and Khatri-Rao products in (B.4) of appendix B, (4.80) can be written:

$$\mathbf{Y}_F = (\mathbf{F}_K^H \mathbf{X}'_L \odot \mathbf{F}_P^H \mathbf{Z}'_L) \mathbf{X}_o \mathbf{Z}_o \mathbf{A} \mathbf{E}_R = \mathbf{B} \mathbf{X}_o \mathbf{Z}_o \mathbf{A} \mathbf{E}_R \tag{4.81}$$

4.5. 2-D Beam-space Matrix Pencil Algorithm

where **B** is a beam-space array manifold matrix. From (4.81), all the matrices to the right of **B** have full row rank. It means that the complex matrix $\mathbf{Y}_F$ and the real matrix **B** share the same column space. The real data matrix can be obtained as

$$\mathbf{Y}_{\text{Re}} = [\text{Re}(\mathbf{Y}_F), \text{Im}(\mathbf{Y}_F)]. \tag{4.82}$$

To reduce the noise effect, the SVD of (3.29) is used, where the left singular vectors $\mathbf{U}_s$ corresponding to the largest singular values should span the column space of **B**. Therefore, $\mathbf{U}_s$ and **B** are related as follows [38]

$$\mathbf{U}_s = \mathbf{BT} \tag{4.83}$$

where **T** is a nonsingular $L \times L$ matrix.

Summary: The summary of the enhanced 2-D BMP for time delays and relative DOAs estimation is described in the following:

Step1: From the estimated CFR of antenna m, find the Hankel matrix $\mathbf{Y}_m$ as in (4.18), and then find the enhanced matrix $\mathbf{Y}_e$ in the Hankel block matrix as in (4.19).

Step2: Apply the conjugate centro-symmetrized version of the DFT matrix as, $\mathbf{Y}_F = \mathbf{F}^H \mathbf{Y}_e$, to get the real data matrix as, $\mathbf{Y}_{\text{Re}} = [\text{Re}(\mathbf{Y}_F), \text{Im}(\mathbf{Y}_F)]$.

Step3: Perform the SVD on $\mathbf{Y}_{\text{Re}}$ as in (3.29) to get the signal subspace, and estimate the number of effective paths $\hat{L}$ using the modified MDL criterion presented in (3.30).

Step4: Determine $\mathbf{U}_s$ as the left singular vectors of **U**, which span the signal subspace and correspond to the largest $\hat{L}$ singular values of $\mathbf{Y}_{\text{Re}}$.

Next, we calculate the selection matrices of the space dimension $\mathbf{\Gamma}_{\mu_1}$ and $\mathbf{\Gamma}_{\mu_2}$, and of the frequency dimension $\mathbf{\Gamma}_{v_1}$ and $\mathbf{\Gamma}_{v_2}$ just once and store them as constant matrices. The ath row of $\mathbf{\Gamma}_1$, $1 \leq a \leq K$, has all its elements equal to zero except the ath and the bth elements as follows [29], [92]

$$x_a = \cos((a-1)\pi/K) \tag{4.84}$$

$$x_b = \begin{cases} \cos(a\pi/K) \,; 1 \leq a \leq K-1, \, b = a+1 \\ (-1)^{K+1} \cos(a\pi/K) \,; a = K, \, b = 1 \end{cases} \tag{4.85}$$

The ath row of $\mathbf{\Gamma}_2$ is expressed in the same way as for $\mathbf{\Gamma}_1$ by replacing cosine functions by sine functions. The selection matrices of the space dimension are

$$\mathbf{\Gamma}_{\mu_1} = \mathbf{\Gamma}_1 \otimes \mathbf{I}_P \tag{4.86}$$

$$\mathbf{\Gamma}_{\mu_2} = \mathbf{\Gamma}_2 \otimes \mathbf{I}_P \tag{4.87}$$

where $\mathbf{I}_P$ is the identity matrix of size $P \times P$. In a similar way, $\mathbf{\Gamma}_3$ and $\mathbf{\Gamma}_4$ are generated similar to $\mathbf{\Gamma}_1$ and $\mathbf{\Gamma}_2$, respectively, where K is replaced by P. The selection matrices of the frequency dimension are then represented

$$\mathbf{\Gamma}_{v_1} = \mathbf{I}_K \otimes \mathbf{\Gamma}_3, \tag{4.88}$$

$$\Gamma_{v_2} = I_K \otimes \Gamma_4. \quad (4.89)$$

Step5: Extracting z_l: The matrix pencil equation of the frequency dimension is

$$\Gamma_{v_1} U_s \Psi_v = \Gamma_{v_2} U_s \Rightarrow \Psi_v = (\Gamma_{v_1} U_s)^\dagger \Gamma_{v_2} U_s, \quad (4.90)$$

and the pencil pair is $(\Gamma_{v_2} U_s, \Gamma_{v_1} U_s)$. The eigenvalues of $\Psi_v = W Z_d W^{-1}$ are computed to get $Z_d = \text{diag}\{\Lambda_l = \tan(v_l/2); l = 1, \ldots, \hat{L}\}$, where $v_l = -2\pi \Delta f \hat{\tau}_l$. The time delays can then be calculated as, $\hat{\tau}_l = -1/\pi\Delta f \times \tan^{-1}(\Lambda_l)$, where $l = 1, \ldots, \hat{L}$.

Step6: Extracting x_l: The matrix pencil equation of the space dimension is

$$\Gamma_{\mu_1} U_s \Psi_\mu = \Gamma_{\mu_2} U_s \Rightarrow \Psi_\mu = (\Gamma_{\mu_1} U_s)^\dagger \Gamma_{\mu_2} U_s, \quad (4.91)$$

and the pencil pair is $(\Gamma_{\mu_2} U_s, \Gamma_{\mu_1} U_s)$. By using the proposed pairing method in Section 4.3.2, it is not necessary to solve an EVD problem. Therefore, compute X_d by premultiplying Ψ_μ with the inverse W'^{-1}, coming from the EVD of z_l poles with the proposed modification, and then postmultiplying with W' to get $X_d = \text{diag}\{\Omega_l = \tan(\mu_l/2); l = 1, \ldots, \hat{L}\}$, where $\mu_l = -2\pi f_c \rho \sin \hat{\theta}_l / c$. After that, calculate the DOA according to $\hat{\theta}_l = \sin^{-1}(-\tan^{-1}(\Omega_l).c/\pi f_c \rho)$, where $l = 1, \ldots, \hat{L}$.

Fig. 4.5 shows a block diagram of the proposed joint propagation time delay and relative DOA estimation algorithms.

4.6 Diversity Techniques using 2-D Matrix Pencil Algorithms

Such as the previous chapter of 1-D MP algorithms, if a number of LTFs are available in the OFDM frame, a number of CFRs can be obtained for each antenna in the antenna array. Therefore, the enhanced matrix Y_e of 2-D MP in (4.19), or the centro-hermitian matrix Y_{ex} of 2-D MP-Ex in (4.32), or the real matrices Y_{Re} in (4.44) and (4.82) of 2-D UMP and 2-D BMP, respectively, can be constructed for each snapshot separately. As

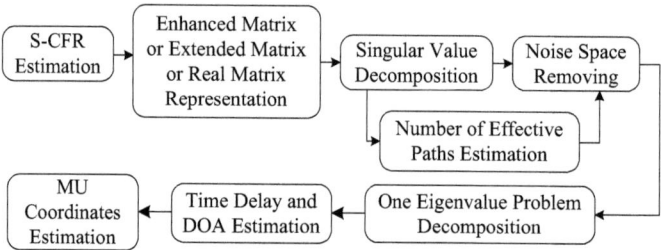

Figure 4.5: Operational flow of the proposed joint propagation time delays and DOAs estimation using 2-D MP algorithms.

we proposed in [54], the multiple snapshot matrix can be formed as in (3.66), $\mathbf{Y}_E = [\mathbf{Y}_0, \mathbf{Y}_1, \ldots, \mathbf{Y}_{q-1}]$, where q is the number of estimated CFRs per each antenna in the antenna array which is equal to the number of LTFs per OFDM frame. Using this type of multiple snapshots is a kind of time diversity. The principle of frequency diversity can also be used to reduce the complexity of high BWs such as in (3.68). If the size of a single snapshot matrix is $a \times b$, it will be in the multiple snapshot matrix $\mathbf{Y}_E$, $a \times qb$, the number of columns is multiplied by q. The same procedure of various 2-D MP algorithms in the previous sections is then applied.

4.7 Computational Complexity

From the previous analysis, the size of the enhanced matrix $\mathbf{Y}_e$ of 2-D MP in (4.19) is $KP \times (M - K + 1)(N - P + 1)$. The centro-hermitian matrix $\mathbf{Y}_{ex}$ of 2-D MP-Ex in (4.32), and the real matrices $\mathbf{Y}_{\text{Re}}$ in (4.44) and (4.82) of 2-D UMP and 2-D BMP, respectively, have a size of $KP \times 2(M-K+1)(N-P+1)$. To measure the computational complexity of various 2-D matrix pencil algorithms, let us investigate the complexity of all steps which are in the following:

- Data matrix transformation in case of using 2-D UMP and 2-D BMP.

- The SVD computation (the singular values and the left singular vectors computation of $\mathbf{Y}_e$ or $\mathbf{Y}_{ex}$ or $\mathbf{Y}_{\text{Re}}$ based on the selected algorithm).

- Matrices of pencil pairs computation.

- The generalized EVD computation of the frequency dimension (the eigenvalues and the eigenvectors of matrix pencil equations computation).

- The DOA poles calculation including the pairing between estimated poles.

Let us now investigate the complexity of each step precisely.

(a) *Data Matrix Transformation*

In the first step, the conventional 2-D MP and 2-D MP-Ex algorithms do not require data transformation processing, where their whole process is based on the complex computation. For 2-D UMP, the UMT transformation is used to get the real data matrix $\mathbf{Y}_{\text{Re}}$ such as in (4.44) by premultiplying $\mathbf{Y}_{ex}$ with $\mathbf{Q}_{K_1}^H$ and then postmultiplying with $\mathbf{Q}_{K_2}$. To do that it requires only scaling and $3 \times KP \times 2(M-K+1)(N-P+1)$ real additions due to the sparse structure of those unitary matrices. For 2-D BMP, the DFT beams are used. The 2-D DFT matrix $\mathbf{F}^H = \mathbf{F}_K^H \otimes \mathbf{F}_P^H$ is calculated once and stored as a constant matrix. To do the initial transformation, the enhanced matrix $\mathbf{Y}_e$ is multiplied by $\mathbf{F}^H$ to get $\mathbf{Y}_F$ as in (4.80), which requires $(KP)^2 \times (M-K+1)(N-P+1)$ complex multiplications.

(b) *The SVD Computation*

In the second step namely the singular values and the left singular vectors computation, the conventional 2-D MP algorithm requires $17K^3P^3/3 + K^2P^2(M - K +$

1)$(N-P+1)$ complex multiplications to decompose $\mathbf{Y}_e$, where $(M-K+1)(N-P+1) > KP$ [83], [36]. For 2-D MP-Ex algorithm, the extended matrix $\mathbf{Y}_{ex}$ is used, where the number of columns increases by a factor of two as in (4.32), consequently, it requires $17K^3P^3/3 + 2K^2P^2 \times (M-K+1)(N-P+1)$ complex multiplications. In 2-D UMP and 2-D BMP algorithms, the size of $\mathbf{Y}_{\text{Re}}$ in (4.44) and (4.82) is like that of $\mathbf{Y}_{ex}$, $KP \times 2(M-K+1)(N-P+1)$, therefore, they require the same number of multiplications for decomposition, but it is in real. It should be noted that the SVD computational complexity of 2-D UMP and 2-D BMP algorithms decreases by a factor of 4 since one complex multiplication requires four real multiplications. In the remaining steps, the required number of multiplications for complex 2-D MP algorithms is also necessary for 2-D UMP and 2-D BMP, the only difference is that it is real valued computations.

(c) *Pencil Pairs Computation*

In the third step namely matrices computation of matrix pencil equations using the signal subspace and the selection matrices in case of real algorithms, the pencil pairs of complex 2-D MP algorithms are $(\mathbf{U}_{s2}, \mathbf{U}_{s1})$ and $(\mathbf{U}_{sp2}, \mathbf{U}_{sp1})$ presented in (4.34) and (4.37) for space and frequency dimensions, respectively, where $\mathbf{U}_{sp} = \mathbf{P}\mathbf{U}_s$. For 2-D UMP algorithm, they are $(\mathbf{K}_{\text{Im}1}\mathbf{U}_s, \mathbf{K}_{\text{Re}1}\mathbf{U}_s)$ and $(\mathbf{K}_{\text{Im}2}\mathbf{U}_{sp}, \mathbf{K}_{\text{Re}2}\mathbf{U}_{sp})$ as in (4.54) and (4.61), where $\mathbf{U}_{sp} = \mathbf{P}'\mathbf{U}_s$ as in (4.59). For 2-D BMP algorithm, they are $(\mathbf{\Gamma}_{v2}\mathbf{U}_s, \mathbf{\Gamma}_{v1}\mathbf{U}_s)$ and $(\mathbf{\Gamma}_{\mu2}\mathbf{U}_s, \mathbf{\Gamma}_{\mu1}\mathbf{U}_s)$ as in (4.90) and (4.91). In case of real 2-D MP algorithms, the selection matrices are multiplied by the signal subspace $\mathbf{U}_s$ of space dimension and $\mathbf{U}_{sp}$ of frequency dimension. Those multiplications require negligible computations, because the whole selection matrices of 2-D UMP and 2-D BMP have the sparse structure, where each row and each column of those selection matrices has no more than two scaling values.

There are two options to compute the matrices of matrix pencil equations based on the EVD method as described in appendix C. For QZ factorization, the matrix pencil pairs should be computed to get the generalized eigenvalue problem equation defined in (C.1) of appendix C. The matrix pencil pairs of 2-D MP are $(\mathbf{U}_{s1}^H\mathbf{U}_{s2}, \mathbf{U}_{s1}^H\mathbf{U}_{s1})$ and $(\mathbf{U}_{sp1}^H\mathbf{U}_{sp2}, \mathbf{U}_{sp1}^H\mathbf{U}_{sp1})$. To compute $\mathbf{U}_{s1}^H\mathbf{U}_{s2}$ and $\mathbf{U}_{s1}^H\mathbf{U}_{s1}$, each requires $\hat{L}^2P(K-1)$ complex multiplications. And to compute $\mathbf{U}_{sp1}^H\mathbf{U}_{sp2}$ and $\mathbf{U}_{sp1}^H\mathbf{U}_{sp1}$, each requires $\hat{L}^2K(P-1)$ complex multiplications. The total number of complex multiplications of this step is $2\hat{L}^2(2KP - K - P)$. Similarly, the matrix pencil pairs of 2-D UMP and 2-D BMP for QZ factorization can be calculated by the same number of multiplications, but it is in real. The matrix pencil pairs of 2-D UMP for QZ factorization are $[(\mathbf{K}_{\text{Re}1}\mathbf{U}_s)^H\mathbf{K}_{\text{Im}1}\mathbf{U}_s, (\mathbf{K}_{\text{Re}1}\mathbf{U}_s)^H\mathbf{K}_{\text{Re}1}\mathbf{U}_s]$ and $[(\mathbf{K}_{\text{Re}2}\mathbf{U}_{sp})^H\mathbf{K}_{\text{Im}2}\mathbf{U}_{sp}, (\mathbf{K}_{\text{Re}2}\mathbf{U}_{sp})^H\mathbf{K}_{\text{Re}2}\mathbf{U}_{sp}]$. The matrix pencil pairs of 2-D BMP for QZ factorization are $[(\mathbf{\Gamma}_{v1}\mathbf{U}_s)^H\mathbf{\Gamma}_{v2}\mathbf{U}_s, (\mathbf{\Gamma}_{v1}\mathbf{U}_s)^H\mathbf{\Gamma}_{v1}\mathbf{U}_s]$ and $[(\mathbf{\Gamma}_{\mu1}\mathbf{U}_s)^H\mathbf{\Gamma}_{\mu2}\mathbf{U}_s, (\mathbf{\Gamma}_{\mu1}\mathbf{U}_s)^H\mathbf{\Gamma}_{\mu1}\mathbf{U}_s]$.

For QR factorization, the matrix pencil pairs should be computed to get the standard eigenvalue problem equation defined in (C.2) of appendix C. To do that the matrices $\mathbf{\Psi}_\mu = \mathbf{U}_{s1}^\dagger\mathbf{U}_{s2}$ and $\mathbf{\Psi}_v = \mathbf{U}_{sp1}^\dagger\mathbf{U}_{sp2}$ given in (4.41) and (4.42), respectively, of the complex 2-D MP algorithms should be computed. The size of those matrices is $\hat{L} \times \hat{L}$. As we know, the superscript $\dagger$ denotes the Moore-Penrose pseudo-inverse,

4.7. Computational Complexity

which is defined for a complex matrix $\mathbf{C}$ as in (3.26), $\mathbf{C}^\dagger = \left(\mathbf{C}^H\mathbf{C}\right)^{-1}\mathbf{C}^H$. Therefore, to calculate $\mathbf{U}_{s1}^H\mathbf{U}_{s1}$ and $\mathbf{U}_{sp1}^H\mathbf{U}_{sp1}$, both require $\hat{L}^2(2KP - K - P)$ complex multiplications. And then to compute $(\mathbf{U}_{s1}^H\mathbf{U}_{s1})^{-1}$ and $(\mathbf{U}_{sp1}^H\mathbf{U}_{sp1})^{-1}$, both require $\frac{2}{3}\hat{L}^3$ complex multiplications [109], [83]. And to calculate $\mathbf{U}_{s1}^H\mathbf{U}_{s2}$ and $\mathbf{U}_{sp1}^H\mathbf{U}_{sp2}$, both require $\hat{L}^2(2KP - K - P)$ complex multiplications. Finally, to construct the whole matrices $\mathbf{U}_{s1}^\dagger\mathbf{U}_{s2}$ and $\mathbf{U}_{sp1}^\dagger\mathbf{U}_{sp2}$, each requires $\hat{L}^3$ complex multiplications. As a result, the total number of complex multiplications is $\frac{8}{3}\hat{L}^3 + 2\hat{L}^2(2KP - K - P)$.

(d) *EVD Computation of Frequency Dimension*

In the fourth step namely EVD computation, the eigenvalues and the eigenvectors should be computed only for frequency dimension based on the proposed pairing method in Section 4.3.2. In case of using QR algorithm, the required number of multiplications to get the Schur form is $15\hat{L}^3$ [109], [83]. To get the eigenvector matrix $\mathbf{W}$, an extra $\hat{L}^3$ multiplications are required, and to get $\mathbf{W}^{-1}$, another $\hat{L}^3$ multiplications are required [109], [83]. The total number of complex multiplications for this step is $17\hat{L}^3$. In case of using QZ algorithm, the required number of multiplications to get the generalized eigenvalues is $5\hat{L}^3$ [36], assuming $\hat{L} \gg 1$.

From the third and fourth steps, it is not recommended to transform the generalized eigenvalue problem into the standard eigenvalue problem. Using the QZ algorithm for EVD is numerically more robust than using the QR algorithm.

(e) *DOA Poles Calculation Including the Pairing*

In the fifth step, to calculate the diagonal elements of $\mathbf{X}_d = \mathbf{W}'^{-1}\mathbf{\Psi}_\mu\mathbf{W}'$, a number of $2\hat{L}^3$ complex multiplications are required.

From the above processing, clearly a single EVD is used instead of two to estimate the required poles of both sets. In addition, the process of pairing has been mitigated. It is worth mentioning that to pair the corresponding eigenvalues using the correlation maximization pairing method in Section 4.3.1, extra multiplications are required $(1/2\hat{L}(\hat{L}+1) - 1)((K-1)(P-1) + \hat{L}(KP+1))$ [36], [37], which can be approximated to $\frac{1}{2}\hat{L}^3 KP$ if $\hat{L} \gg 1$, $K \gg 1$, and $P \gg 1$. In addition, by using the priori information of estimated time delays, the problem of repeated poles has been mitigated, which requires an additional EVD for each repeated pole as proposed in [115] for azimuth and elevation angles estimation using the URA.

In this work, it is preferred to select the pencil parameter values of P and K to be $N/3$ and $M/3$, respectively, to reduce the complexity and to stay in the appropriate range. Table 4.1 shows the selected values of pencil parameter K with respect to the number of antennas per each BS. Table 4.2 shows the selected values of pencil parameter P with respect to the length of the interpolated CFR based on the selected bandwidth of 802.11ac. In case of using the multiple snapshot principle for temporal diversity presented in Section 4.6 as in (3.66), the number of columns is multiplied by the number of snapshots q. In case of using the multiple snapshot principle for frequency diversity as in (3.68), the new pencil value P of the smaller bandwidth from Table 4.2 is used to generate the enhanced matrices of the left and right parts of the large bandwidth, and then they are combined such as in (3.68).

82 Chapter 4. Joint Time Delay and DOA Estimation using 2-D Matrix Pencil Algorithms

From the above, it should be noted that the most computationally intensive step is to estimate the signal subspace using the SVD in all 2-D matrix pencil algorithms. Therefore, to show the comparison of the computational complexity of the initial transformation and SVD of various 2-D matrix pencil algorithms, let us assume the number of antennas M is 6, hence, K is 2, and the pencil parameter P is given in Table 4.2 for all 802.11ac bandwidths. The comparison is shown in Fig. 4.6. The 2-D UMP is the best regarding the complexity. Table 4.3 shows the complexity ratio of 2-D UMP to the other 2-D matrix pencil algorithms. The complexity ratios of 2-D UMP to 2-D MP-Ex, 2-D MP, and to 2-D BMP are 0.25, 0.368, and 0.438, respectively.

4.8 Computational Complexity Comparison of 1-D and 2-D Matrix Pencil Algorithms

From the previous chapter of 1-D MP algorithms, we have proposed that if a number of antennas are available in the BS, the 1-D MP algorithms can be used, where the measured CFRs from all antennas per each BS can be treated as a number of snapshots, in another way, as a kind of spatial diversity. In this section, we need to present the computational complexity of 1-D MP algorithms with spatial diversity to estimate only time delays compared with that of 2-D MP algorithms to estimate both time delays and relative DOAs. To show the computational complexity comparison between both principles, let us present the comparison between 1-D UMP-Ex and 2-D UMP, where they are using the extended matrix and they achieve the lowest computational Complexity. To do that let us assume the number of antennas $M = 6$, consequence, the number of snapshots for spatial diversity using 1-D UMP-Ex is $q = 6$. The pencil parameter P of 1-D UMP-Ex is given in Table 3.5 for all bandwidths of 802.11ac and the pencil parameters K and P of 2-D UMP are given in Tables 4.1 and 4.2, respectively. Fig. 4.7 shows a comparison between the computational complexity of the SVD of both 1-D UMP-Ex and 2-D UMP algorithms regarding 802.11ac bandwidths. From Fig. 4.7, it is obvious that the computational complexity of 1-D UMP-Ex is very low compared to that of 2-D UMP. For $2 \leq M \leq 6$, the complexity ratio of 1-D UMP-Ex to 2-D UMP is around $\varrho = 0.237$ and in case of using frequency diversity it is $\varrho = 0.261$. For $7 \leq M \leq 8$, $\varrho \approx 0.1$. The question now is what is the performance ratio between using 1-D and 2-D matrix pencil algorithms for wireless positioning. Furthermore, is there a mismatch between the complexity and the performance of each of them? The answer of those questions will be given in the measurement chapter.

In Chapter 3, the time delays of a wireless multipath channel have been estimated, and in this chapter, the time delays and the relative DOAs of a wireless multipath channel have been estimated. After estimating the required observations, the next step is to estimate

Table 4.1: The selected values of pencil parameter K with respect to the number of antennas.

Number of antennas	2	3	4	5	6	7	8
K	2	2	2	2	2	3	3

4.8. Computational Complexity Comparison of 1-D and 2-D Matrix Pencil Algorithms

Table 4.2: The selected values of pencil parameter P versus the CFR length.

Bandwidth (MHz)	20	40	80	160
FFT / IFFT order	64	128	256	512
$N = N_p + N_{dc}$	57	117	245	501
P	19	39	82	167

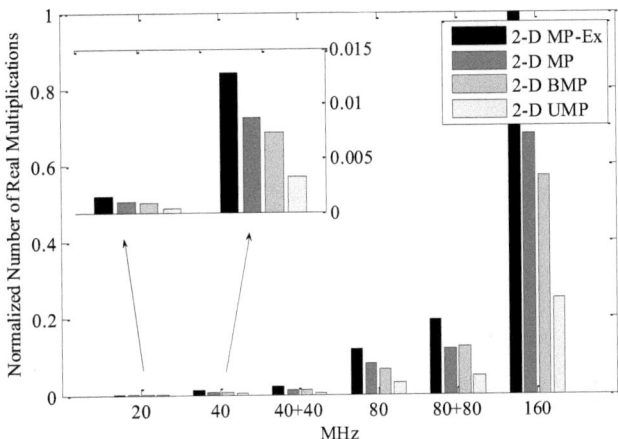

Figure 4.6: Comparison of computational complexity of the initial transformation and SVD of various 2-D MP algorithms.

Table 4.3: The complexity ratio of 2-D UMP to the other 2-D matrix pencil algorithms.

Bandwidth (MHz)	2-D UMP to 2-D MP-Ex	2-D UMP to 2-D MP	2-D UMP to 2-D BMP
20	0.25	0.3688	0.4370
40	0.25	0.3680	0.4381
40+40	0.25	0.4103	0.3902
80	0.25	0.3672	0.4393
80+80	0.25	0.4096	0.3909
160	0.25	0.3674	0.4390

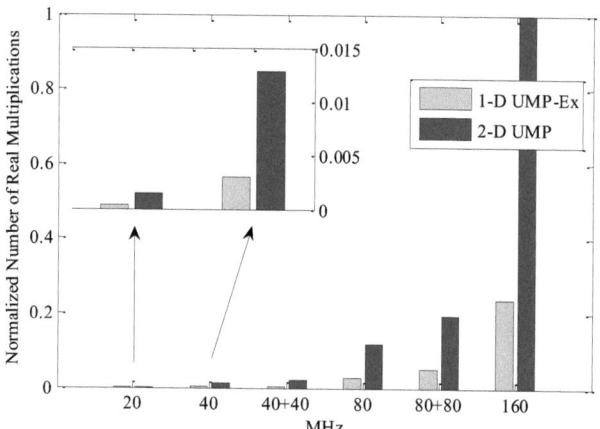

Figure 4.7: Comparison of computational complexity of the SVD between 1-D UMP-Ex and 2-D UMP algorithms.

the MU coordinates using TDOA observations, or hybrid TDOA and DOA observations. Chapter 5 will present some useful studies and the proposed estimators for MU position estimation. The effect of the fundamental parameters of the wireless positioning system on the performance will also be presented.

CHAPTER 5
Mobile Unit Position Estimation Based on TDOA and DOA Measurements

There are two options to combine TDOA and DOA estimates for MU position estimation. The first option is to estimate the coordinates of MU using TDOA and DOA measurements individually, and then combine the estimated positions to get the final position, where each result should be given an appropriate weight due to the different variance error. The second option is to process the TDOA and DOA measurements together to estimate the MU position. In the following, some useful studies to estimate the MU position by using TDOA, DOA, and hybrid TDOA and DOA measurements will be presented in Sections 5.1, 5.2, and 5.3, respectively. The proposed TDOA, and hybrid TDOA and DOA estimators will be presented in Section 5.4. After that the TDOA and DOA estimation error variances will be derived to show the effect of the fundamental parameters of the wireless indoor positioning system in Section 5.5.

5.1 Estimators Based on TDOA Measurements

The geometric model for estimating the position coordinates using TDOA is the intersection of hyperbolas in 2-D and the intersection of hyperboloid in 3-D as shown in Fig. 2.15. Fig. 5.1 shows a simple demo, where four BSs are used. The coordinates of these BSs from 1 to 4 are (x_1, y_1), (x_2, y_2), (x_3, y_3), and (x_4, y_4), respectively, which are known to the system. The position of the MU is (x, y) (this is the parameter to be estimated). The TDOA between receivers i and j can be computed as

$$\Delta t_{ij} = t_i - t_j; i, j = 1, 2, \ldots, I; i \neq j \qquad (5.1)$$

where I is the number of fixed BSs. If BS number one is taken as a reference, the TDOA between BS number i and the reference BS is

$$\Delta t_{i1} = t_i - t_1; i = 2, \ldots, I \qquad (5.2)$$

The equations of the hyperbolas that define the position of the MU are

$$d_{i1} = c.\Delta t_{i1} = \hat{d}_i - \hat{d}_1 = \sqrt{(\hat{x} - x_i)^2 + (\hat{y} - y_i)^2} - \sqrt{(\hat{x} - x_1)^2 + (\hat{y} - y_1)^2} \qquad (5.3)$$

where d_{i1} is the TDOA between BS number i and the reference BS times the speed of light, c. The solution of the above set of non-linear equations gives $\hat{x}, \hat{y}$ and $\hat{d}_1$. However, it is difficult to solve those non-linear equations. Therefore, linearizing (5.3) is the possible way to solve it. In the following, different positioning estimators based on the TDOA observations to estimate MU location $(\hat{x}, \hat{y})$ will be introduced.

86 Chapter 5. Mobile Unit Position Estimation Based on TDOA and DOA Measurements

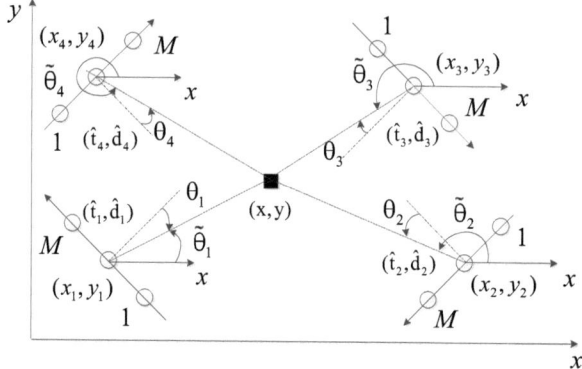

Figure 5.1: Deployment of BSs for wireless positioning based on TDOA and DOA.

5.1.1 Iterative Least Square Estimator

The linearized LS estimator is a ML estimator that uses Taylor series expansion to linearize (5.3) [42]. This estimator starts with an initial guess for the MU position (x_0, y_0), and then calculates the position deviation for each iteration. In case of wireless indoor positioning, (x_0, y_0) can be selected the center of the interested area. In general, the iterative LS estimator computes position deviations using Taylor series expansion as follows:

$$\hat{\mathbf{z}} = \begin{bmatrix} \hat{x} \\ \hat{y} \end{bmatrix} = \begin{bmatrix} x_0 \\ y_0 \end{bmatrix} + (\mathbf{G}^T \mathbf{Q}^{-1} \mathbf{G})^{-1} \mathbf{G}^T \mathbf{Q}^{-1} \mathbf{h} \qquad (5.4)$$

where **G** is the designed matrix and defined by

$$\mathbf{G} = \begin{bmatrix} (x_1 - x_0)/d_1 - (x_2 - x_0)/d_2 & (y_1 - y_0)/d_1 - (y_2 - y_0)/d_2 \\ (x_1 - x_0)/d_1 - (x_3 - x_0)/d_3 & (y_1 - y_0)/d_1 - (y_3 - y_0)/d_3 \\ \vdots & \vdots \\ (x_1 - x_0)/d_1 - (x_I - x_0)/d_I & (y_1 - y_0)/d_1 - (y_I - y_0)/d_I \end{bmatrix}, \qquad (5.5)$$

$$\mathbf{h} = \begin{bmatrix} d_{21} - (d_2 - d_1) \\ d_{31} - (d_3 - d_1) \\ \vdots \\ d_{I1} - (d_I - d_1) \end{bmatrix}, \qquad (5.6)$$

where the values of $d_i; i = 1, 2, ..., I$ are computed using the initial guess $\mathbf{z}_0 = [x_0, y_0]^T$. The new estimated coordinates $(\hat{x}_0, \hat{y}_0)$ are used as an initial estimate in (5.4) for the next iteration. And **Q** is the covariance matrix converted from its original TDOA form into the corresponding distance form by multiplication by c^2. The whole process should be repeated until the deviations between successive estimated values of $\hat{\mathbf{z}} = [\hat{x}, \hat{y}]^T$ are sufficiently small. Although the number of independent TDOA values obtainable from I

5.1. Estimators Based on TDOA Measurements

BSs is $I-1$, we can use in a noisy environment additional pairs of measurements that are not independent, since the noise that is not correlated between those pairs gives them a degree of independence [89]. Therefore, more pairs can also be used in the iterative LS estimator. For example, if we have four BSs, it is possible to take these pairs of BSs to calculate TDOAs: (1,2), (1,3), (1,4), (2,3), (2,4), and (3,4). The MU position can then be estimated iteratively using (5.4), where extra rows in G and h matrices should be added to take into account the new pairs. In this case, it has been found from the experimental results that the covariance matrix should be neglected, because it is a singular matrix, and cannot be inverted; its determinant is approximately zero.

The drawback of iterative LS is that it is a computationally intensive method if the starting point is not close enough. In addition, it has a convergence problem, convergence is not guaranteed [43].

5.1.2 Divide-and-Conquer based on TDOA Estimator

Divide-and-Conquer estimator has been presented in [44] and [45], which can achieve a good performance at high SNR. The principle of DAC estimator is to split the observations to small sets each having a size equal to the number of unknowns. The MU coordinates are then estimated for each set, $(\hat{z}_1, \hat{z}_2, ..., \hat{z}_K)$, where K is the number of sets, and $\hat{z}_k = [\hat{x}_k, \hat{y}_k]^T$. The final estimate can then be obtained by combining those estimates. Here, the basic principle of DAC will be presented with some modifications. The proposed LS estimator, which will be presented in Section 5.4.1, is used to estimate the MU coordinates of each set. Then, the estimations will be combined by giving a weight to each estimation using the proposed method in [44]. If we have four BSs such as in Fig. 5.1, the proposed LS estimator is used once for each set as: set1 (1,2,3), set2 (2,3,4), set3 (3,4,1), and set4 (4,1,2). As it has been explained in the previous section, we can use in a noisy environment additional pairs of measurements that are not independent. The final ML estimate $\hat{z}_{ML}$ is the LS estimate of

$$\hat{z}_{ML} = arg \min_{\mathbf{z}} \sum_{k=1}^{K} (\mathbf{d}_k - \mathbf{f}_k(\mathbf{z}))^T \mathbf{Q}_k (\mathbf{d}_k - \mathbf{f}_k(\mathbf{z})) \qquad (5.7)$$

where $\mathbf{d}_k = [d_{21}, d_{31}]^T$ is a vector containing the measured TDOAs of set k multiplied by c as shown in (5.3), $\mathbf{Q}_k$ is the covariance matrix of set k converted to distance form, and $\mathbf{f}_k(\mathbf{z}) = [f_2(\mathbf{z}), f_3(\mathbf{z})]^T$ is a vector containing the well known non-linear hyperbolic functions shown in (5.3) as [45]

$$\begin{aligned} f_i(\mathbf{z}) &= \|\mathbf{z} - \mathbf{z}_i\| - \|\mathbf{z} - \mathbf{z}_1\| \\ &= \sqrt{(x-x_i)^2 + (y-y_i)^2} - \sqrt{(x-x_1)^2 + (y-y_1)^2} \end{aligned} \qquad (5.8)$$

where $\mathbf{z}_i = [x_i, y_i]^T$ is the coordinates of BS number i, $\mathbf{z}_1 = [x_1, y_1]^T$ is the coordinates of the reference BS of set k, and $\mathbf{z} = [x, y]^T$ is the coordinates of MU, which should be estimated. From [44], the ML position estimate $\hat{z}_{ML}$ shown in (5.7) can be approximated as follows

$$\hat{z}_{ML} \approx \sum_{k=1}^{K} \mathbf{W}_k \hat{z}_k \qquad (5.9)$$

where the weighting vector $\mathbf{W}_k$ of $\hat{\mathbf{z}}_k$ that has been estimated from set k is

$$\mathbf{W}_k = [\sum_{k=1}^{K} \mathbf{C}_k]^{-1} \mathbf{C}_k \qquad (5.10)$$

where $\mathbf{C}_k$ is given by

$$\mathbf{C}_k = \mathbf{G}_k^T \mathbf{Q}_k^{-1} \mathbf{G}_k \qquad (5.11)$$

where $\mathbf{G}_k = \nabla \mathbf{f}_k(\mathbf{z})$ is the gradient of $\mathbf{f}_k(\mathbf{z})$ with respect to $\mathbf{z}$. In case of 2-D wireless positioning, each set should have three BSs to estimate $\mathbf{z}_k$, therefore, $\mathbf{G}_k$ could be written as

$$\mathbf{G}_k = \begin{bmatrix} \nabla f_2(\mathbf{z}) & \nabla f_3(\mathbf{z}) \end{bmatrix}^T. \qquad (5.12)$$

The gradient of $f_i(\mathbf{z})$ is

$$\nabla f_i(\mathbf{z}) = \begin{bmatrix} \frac{\partial f_i(\mathbf{z})}{\partial x} & \frac{\partial f_i(\mathbf{z})}{\partial y} \end{bmatrix} \qquad (5.13)$$

where

$$\frac{\partial f_i(\mathbf{z})}{\partial x} = \frac{x - x_i}{\sqrt{(x - x_i)^2 + (y - y_i)^2}} - \frac{x - x_1}{\sqrt{(x - x_1)^2 + (y - y_1)^2}}, \qquad (5.14)$$

and

$$\frac{\partial f_i(\mathbf{z})}{\partial y} = \frac{y - y_i}{\sqrt{(x - x_i)^2 + (y - y_i)^2}} - \frac{y - y_1}{\sqrt{(x - x_1)^2 + (y - y_1)^2}}. \qquad (5.15)$$

To summarize that first the intersections between hyperbolas defined by pairs of TDOA measurements are estimated using the proposed LS estimator in Section 5.4.1. Then, the final position estimate is calculated using (5.9) by combining the partial position estimates, where each estimate has been given a weighting value calculated from (5.10).

The drawback of this estimator is that sets should have large enough Fisher information [45], [43]. Therefore, optimum performance can be achieved only if the noise is small enough.

5.1.3 Chan Estimator

The principle of Chan estimator presented in [43] will be explained in the following. From Fig. 5.1, the squared distance between the MU and BS number i is

$$d_i^2 = (x_i - x)^2 + (y_i - y)^2 = K_i - 2x_i x - 2y_i y + x^2 + y^2 \qquad (5.16)$$

where $K_i = x_i^2 + y_i^2$. The set of non-linear equations whose solution gives MU position has been defined in (5.3). The first step of Chan estimator is to assume that there is no relationship between x, y, and d_1. Another approximation to make the problem solvable is to assume that the MU has equal distances to all BSs. In another way, it is far from system array, i.e. $d_1 = d_2 = ... = d_I$. The initial estimate of $\mathbf{z}$ can then be estimated by

$$\hat{\mathbf{z}}_a = [\hat{x}\ \hat{y}\ \hat{d}_1]^T = (\mathbf{G}^T \mathbf{Q}^{-1} \mathbf{G})^{-1} \mathbf{G}^T \mathbf{Q}^{-1} \mathbf{h} \qquad (5.17)$$

5.1. Estimators Based on TDOA Measurements

where **Q** is the covariance matrix converted from its original TDOA form into the corresponding distance form. The matrix **G** is defined as

$$\mathbf{G} = \begin{bmatrix} (x_2 - x_1) & (y_2 - y_1) & d_{21} \\ (x_3 - x_1) & (y_3 - y_1) & d_{31} \\ \vdots & \vdots & \vdots \\ (x_I - x_1) & (y_I - y_1) & d_{I1} \end{bmatrix}. \tag{5.18}$$

And the matrix **h** is defined as

$$\mathbf{h} = \begin{bmatrix} d_{21}^2 - K_2 + K_1 \\ d_{31}^2 - K_3 + K_1 \\ \vdots \\ d_{I1}^2 - K_I + K_1 \end{bmatrix}. \tag{5.19}$$

After obtaining the initial estimate $\hat{\mathbf{z}}_a$ using (5.17), a diagonal matrix $\mathbf{B} = \text{diag}(d_2, d_3, ..., d_I)$ can be easily calculated from $\hat{\mathbf{z}}_a$. The covariance matrix is then given by

$$\mathbf{\Psi} = \mathbf{BQB}. \tag{5.20}$$

In case of wireless indoor positioning, the MU is close, and it has different distances to BSs. The second step is to take that into account, hence, the MU position can be estimated as

$$\hat{\mathbf{z}}_b = (\mathbf{G}^T \mathbf{\Psi}^{-1} \mathbf{G})^{-1} \mathbf{G}^T \mathbf{\Psi}^{-1} \mathbf{h}. \tag{5.21}$$

Eq. (5.21) can also be iterated to provide a good estimate. The previous solution of $\hat{\mathbf{z}}_b = [\hat{x}, \hat{y}, \hat{d}_1]^T$ assumes that x, y, and d_1 are independent, where they are related by (5.16). The final step of Chan method is to incorporate this relationship to give an improved estimate. The derivation here is omitted and can be seen in [43]. By using the coordinates of BS one as a reference, the deviations can be calculated as

$$\Delta \hat{\mathbf{z}} = (\mathbf{G}_a^T \mathbf{\Psi}_a^{-1} \mathbf{G}_a)^{-1} \mathbf{G}_a^T \mathbf{\Psi}_a^{-1} \mathbf{h}_a \tag{5.22}$$

where

$$\mathbf{h}_a = \begin{bmatrix} (\hat{x} - x_1)^2 \\ (\hat{y} - y_1)^2 \\ (\hat{d}_1)^2 \end{bmatrix}, \tag{5.23}$$

$$\mathbf{G}_a = \begin{bmatrix} 1 & 0 \\ 0 & 1 \\ 1 & 1 \end{bmatrix}, \tag{5.24}$$

and

$$\mathbf{\Psi}_a = 4\mathbf{B}_a \text{cov}(\hat{\mathbf{z}}_b) \mathbf{B}_a, \tag{5.25}$$

where

$$\mathbf{B}_a = \text{diag}(\hat{x} - x_1, \hat{y} - y_1, \hat{d}_1) \tag{5.26}$$

where d_1 in (5.26) is calculated using $\hat{x}$ and $\hat{y}$. The covariance matrix of $\hat{\mathbf{z}}_b$ is defined as

$$\text{cov}(\hat{\mathbf{z}}_b) = (\mathbf{G}^T \mathbf{\Psi}^{-1} \mathbf{G})^{-1}. \tag{5.27}$$

The final position estimate is then obtained from $\Delta \hat{\mathbf{z}}$ as

$$\hat{\mathbf{z}} = \sqrt{\Delta \hat{\mathbf{z}}} + [x_1, y_1]^T \tag{5.28}$$

or

$$\hat{\mathbf{z}} = -\sqrt{\Delta \hat{\mathbf{z}}} + [x_1, y_1]^T. \tag{5.29}$$

The final solution is selected to lie in the region of interest. This estimator offers a computational advantage over the iterative LS estimator and eliminates the convergence problem.

The drawbacks of Chan estimator are that it is based on the assumption that the noises in the TDOA measurements are small. It will be observed clearly in the measurements analysis. It assumes initially the MU coordinates and the reference distance to the reference BS are independent, and the MU has equal distances to all BSs, in another way, the MU is located in the far field from the system. It needs three steps or more to get the final position estimate of MU.

5.2 Estimators Based on DOA Measurements

For DOA systems, the location and distances are found by triangulation. An example is shown in Fig. 5.1. It should be noted that the DOA support is limited to one semi-plane, $\hat{\theta} \in [-\pi/2, \pi/2]$, which is derived from the π ambiguity of subspace methods using ULAs with isotropic antennas [108]. The reference antenna should be defined for each antenna array. For example in Fig. 5.1, for BS number 1, the array axis goes from the reference antenna to antenna number M. We recommend antenna numbering presented in Fig. 5.1, where it seems that antenna lines build a closed loop around the interested area. The DOA estimation at BS number i is defined with respect to the specific reference system of BS number i. From Fig. 2.2 and 5.1, the estimated angle $\hat{\theta}$ represents the angle between the orthogonal line to the array axis and the impinging wave. The sign of the estimated DOA depends on the impinging wave direction, where it is positive if the impinging wave comes from the right side of the orthogonal line, and it is negative if the impinging wave comes from the left side of this orthogonal line. Therefore, the orientation of all antenna arrays in the positioning system should be known precisely. As an example in Fig. 5.1, θ_1 and θ_3 are positive angles, while θ_2 and θ_4 are negative angles. A common reference system should be used to define $\tilde{\theta}_i, i = 1, ..., I$, with respect to the positive direction of the x axis. If four BSs are located in the XY plane, the estimated DOAs at the four BSs are $\hat{\theta}_1$, $\hat{\theta}_2$, $\hat{\theta}_3$, and $\hat{\theta}_4$. The estimated DOA of each antenna array should be calibrated to the whole system, for example in Fig. 5.1, the DOAs in degrees according to the common reference system are $\tilde{\theta}_1 = 45 - \hat{\theta}_1$, $\tilde{\theta}_2 = 135 - \hat{\theta}_2$, $\tilde{\theta}_3 = -135 - \hat{\theta}_3$, and $\tilde{\theta}_4 = -45 - \hat{\theta}_4$. Although two BSs are enough to localize the MU using DOA methods as it is shown in Fig. 2.12, more than two BSs are necessary, because the DOAs cannot be measured exactly and to mitigate the situation in which the BSs are located on a straight line passing through the MU. In fact, the minimum number of BSs should be three [108].

First, let us show how to find the coordinates of the MU using two BSs, for example BS one and two in Fig. 5.1. By using the triangulation relationships, the non-linear equations can be written as

$$\tan(\tilde{\theta}_i) = (\hat{y} - y_i)/(\hat{x} - x_i); i = 1, \ldots, I. \tag{5.30}$$

5.2. Estimators Based on DOA Measurements

For $I = 2$, (5.30) can be rewritten as

$$\hat{y} = (\hat{x} - x_1)\tan(\tilde{\theta}_1) + y_1, \tag{5.31}$$

$$\hat{x} = (\hat{y} - y_2)/\tan(\tilde{\theta}_2) + x_2. \tag{5.32}$$

By replacing (5.31) into (5.32), $\hat{x}$ can be obtained by

$$\hat{x} = \frac{y_1 - y_2 - x_1\tan(\tilde{\theta}_1) + x_2\tan(\tilde{\theta}_2)}{\tan(\tilde{\theta}_2) - \tan(\tilde{\theta}_1)}. \tag{5.33}$$

By replacing (5.33) into (5.31), $\hat{y}$ can be obtained by

$$\hat{y} = \frac{y_1\tan(\tilde{\theta}_2) - y_2\tan(\tilde{\theta}_1) + x_2\tan(\tilde{\theta}_2)\tan(\tilde{\theta}_1) - x_1\tan(\tilde{\theta}_2)\tan(\tilde{\theta}_1)}{\tan(\tilde{\theta}_2) - \tan(\tilde{\theta}_1)}. \tag{5.34}$$

By using (5.33) and (5.34), the MU position $(\hat{x}, \hat{y})$ can be estimated for each two BSs. It is worth mentioning that the estimation accuracy using DOA observations depends on the position of MU with respect to the BSs. The highest accuracy could be achieved if the positions of MU and BSs form an acute triangle (all angles are less than 90°) [6]. In the following, different positioning methods based on the DOA observations will be introduced to estimate the MU location $(\hat{x}, \hat{y})$.

5.2.1 Lines Intersection Estimator

The simplest method is to estimate the MU coordinates $\mathbf{z} = [x, y]^T$ using (5.33) and (5.34), which result from the intersection of possible lines. For example, if we have four BSs as shown in Fig. 5.1, $\mathbf{z}$ can be estimated using the following sets of BSs: (1,2), (1,4), (2,3), and (3,4). It should be noted that the worst cases of BSs combinations have been mitigated such as (1,3) or (2,4), where those BSs are approximately located in the front of each other, consequence, the intersection of their lines could not be available if their DOA measurements are noisy. The final estimate of MU position can then be obtained using the median of the estimated values.

5.2.2 Least Square Estimator

A simple linear LS position estimator can be obtained by using the triangulation relationships presented in (5.30), which can be written as:

$$\hat{x}\tan(\tilde{\theta}_i) - \hat{y} = x_i\tan(\tilde{\theta}_i) - y_i; i = 1, \ldots, I \tag{5.35}$$

which can be written in a matrix form as:

$$\mathbf{G}.\hat{\mathbf{z}} = \mathbf{b} \tag{5.36}$$

where

$$\mathbf{G} = \begin{bmatrix} \tan(\tilde{\theta}_1) & -1 \\ \vdots & \vdots \\ \tan(\tilde{\theta}_I) & -1 \end{bmatrix}, \tag{5.37}$$

$$\mathbf{b} = \begin{bmatrix} x_1 \tan(\tilde{\theta}_1) - y_1 \\ \vdots \\ x_I \tan(\tilde{\theta}_I) - y_I \end{bmatrix}. \quad (5.38)$$

The solution of $\hat{\mathbf{z}} = [\hat{x}, \hat{y}]^T$ is then given by

$$\hat{\mathbf{z}} = \mathbf{G}^\dagger \mathbf{b} = (\mathbf{G}^T \mathbf{G})^{-1} \mathbf{G}^T \mathbf{b}. \quad (5.39)$$

The drawback of this estimator is that it does not include any information about the variance of the DOA measurements. Therefore, it can be improved by giving a weight to the LS estimator as

$$\hat{\mathbf{z}} = (\mathbf{G}^T \mathbf{d}^{-1} \mathbf{Q}_{DOA}^{-1} \mathbf{G})^{-1} \mathbf{G}^T \mathbf{d}^{-1} \mathbf{Q}_{DOA}^{-1} \mathbf{b} \quad (5.40)$$

where

$$\mathbf{d} = \mathrm{diag}(d_1^2, ..., d_I^2) \quad (5.41)$$

which represents the distances between the BSs and the MU, which are unknown. However, the dependency of the estimator on $\{d_i\}$ is weak as presented in [47], [108]. Therefore, they can be roughly calculated using the center of the interested area, or estimated from the RSS, or simply replaced by an identity matrix. And $\mathbf{Q}_{DOA}$ is the variance matrix which can be represented as follows:

$$\mathbf{Q}_{DOA} = \begin{bmatrix} \sigma_1^2 & \cdots & 0 \\ \vdots & \ddots & \vdots \\ 0 & \cdots & \sigma_I^2 \end{bmatrix} \quad (5.42)$$

where $\{\sigma_i^2; i = 1, \ldots, I\}$ are the DOA variances, which have been calculated from the DOA measurements of each BS.

5.2.3 Divide-and-Conquer based on DOA Estimator

The principle of DAC estimator based on DOA observations is like that of DAC based on TDOA observations presented in Section 5.1.2 [46]. If we have a number of BSs I, which can measure I DOAs of a single MU. The measured DOA at BS number i can be written as

$$\tilde{\theta}_i = f_i(\mathbf{z}) + w_i \quad (5.43)$$

where w_i is a zero mean uncorrelated Gaussian noise with known variance σ_{DOA}^2. From (5.30), the non-linear relationship between the DOA measurements $f_i(\mathbf{z})$ and the MU position can be expressed as

$$f_i(\mathbf{z}) = \tan^{-1}\left(\frac{y - y_i}{x - x_i}\right). \quad (5.44)$$

Eq. (5.44) can be represented in a vector form by including all I DOA measurements as

$$\boldsymbol{\theta} = \mathbf{f}(\mathbf{z}) + \mathbf{w} \quad (5.45)$$

where $\boldsymbol{\theta} = [\tilde{\theta}_1, ..., \tilde{\theta}_I]^T$, $\mathbf{f}(\mathbf{z}) = [f_1(\mathbf{z}), ..., f_I(\mathbf{z})]^T$, and $\mathbf{w} = [w_1, ..., w_I]^T$. The principle of DAC estimator is to split the observations to small sets each having a size equal to the

5.2. Estimators Based on DOA Measurements

number of unknowns. The MU coordinates are then estimated for each pair $(\hat{\mathbf{z}}_1, \hat{\mathbf{z}}_2, ..., \hat{\mathbf{z}}_K)$, where K is the number of pairs, and $\hat{\mathbf{z}}_k = [\hat{x}_k, \hat{y}_k]^T$. The final estimate can then be obtained by combining those ML estimates. From [44], the ML position estimate $\hat{\mathbf{z}}_{ML}$ could be estimated using (5.9), where the weighting vector $\mathbf{W}_k$ of $\hat{\mathbf{z}}_k$, which has been estimated from set k, is defined as

$$\mathbf{W}_k = [\sum_{k=1}^{K} \mathbf{C}'_k]^{-1} \mathbf{C}'_k \qquad (5.46)$$

where $\mathbf{C}'_k$ is given by

$$\mathbf{C}'_k = \mathbf{G}'_k{}^T \mathbf{d}^{-1} \mathbf{Q}'_k{}^{-1} \mathbf{G}'_k \qquad (5.47)$$

where $\mathbf{Q}'_k = \text{diag}(\sigma_1^2, \sigma_2^2)$ is the variance matrix of DOA measurements of pair k, and $\mathbf{d} = \text{diag}(d_1^2, d_2^2)$, where d_1 and d_2 are the distances between the MU and the BSs of pair k, which are unknown and can be calculated to the center of the interested area. And $\mathbf{G}'_k$ is the Jacobian matrix of pair k. It represents the gradient of $\mathbf{f}_k(\mathbf{z}) = [f_1(\mathbf{z}), f_2(\mathbf{z})]^T$ with respect to $\hat{\mathbf{z}}_k$ which can be obtained as [46]

$$\mathbf{G}'_k(\hat{\mathbf{z}}_k) = \nabla \mathbf{f}_k(\hat{\mathbf{z}}_k) = \begin{bmatrix} \mathbf{g}_1(\hat{\mathbf{z}}_k) & \mathbf{g}_2(\hat{\mathbf{z}}_k) \end{bmatrix}^T \qquad (5.48)$$

where

$$\mathbf{g}_i(\hat{\mathbf{z}}_k) = \frac{1}{\|\hat{\mathbf{z}}_k - \mathbf{z}_i\|} \begin{bmatrix} -\sin(f_i(\hat{\mathbf{z}}_k)) & \cos(f_i(\hat{\mathbf{z}}_k)) \end{bmatrix}. \qquad (5.49)$$

If we have I BSs, it has been proposed in [46] that the total number of pairs is $I/2$, for example, the interesting pairs are $(2i, 2i-1)$, where $i = 1, 2, ..., I/2$. The drawback of this method is that if the number of BSs is odd, the DOA estimate of the remaining BS cannot be utilized to improve the position estimation accuracy. To improve the previous method, we can propose the following order for the existing BSs to find the lines intersection between the selected pairs. The following interesting pairs are used $(i, i + 1)$, when $i = 1, 2, ..., I - 1$, and $(i, 1)$, when $i = I$. For example, if $I = 4$ as shown in Fig. 5.1, the interesting pairs are (1,2), (2,3), (3,4), and (4,1). For different distribution of BSs, all pairs should be used without those pairs whose BSs are located in the front of each other, because for a small noise in their DOA measurements, their lines may not intersect.

Such as the previous, the drawback of DAC estimator based on DOA is that a good performance can be achieved only when the noise is small enough. For example in our experimental results, the results of some pairs are empty, there is no intersection between their lines.

5.2.4 Stansfield-Least Square Estimator

To solve the problem of partitioning, a non-iterative closed form has been proposed based on the DOA measurements in [47], [117]. It is based on an approximation of the ML problem to linearize the non-linear ML problem [118]. By using the triangulation relationships, the non-linear equations presented in (5.30) can be written as

$$-\hat{x}\sin(\tilde{\theta}_i) + \hat{y}\cos(\tilde{\theta}_i) = x_i \sin(\tilde{\theta}_i) + y_i \cos(\tilde{\theta}_i) \qquad (5.50)$$

Based on the LS principle in the absence of noise, the proposed position estimate is

$$\hat{\mathbf{z}} = (\mathbf{G}^T \mathbf{G})^{-1} \mathbf{G}^T \mathbf{b} \qquad (5.51)$$

where

$$\mathbf{G}(\tilde{\theta}) = \begin{bmatrix} -\sin(\tilde{\theta}_1) & \cos(\tilde{\theta}_1) \\ \vdots & \vdots \\ -\sin(\tilde{\theta}_I) & \cos(\tilde{\theta}_I) \end{bmatrix}, \quad (5.52)$$

and

$$\mathbf{b}(\tilde{\theta}) = \begin{bmatrix} -x_1\sin(\tilde{\theta}_1) + y_1\cos(\tilde{\theta}_1) \\ \vdots \\ -x_I\sin(\tilde{\theta}_I) + y_I\cos(\tilde{\theta}_I) \end{bmatrix}. \quad (5.53)$$

The drawback of this estimator is that it does not include any information about the variance of the DOA measurements. Therefore, it can be improved by including the variance matrix as

$$\hat{\mathbf{z}} = (\mathbf{G}^T\mathbf{d}^{-1}\mathbf{Q}_{DOA}^{-1}\mathbf{G})^{-1}\mathbf{G}^T\mathbf{d}^{-1}\mathbf{Q}_{DOA}^{-1}\mathbf{b} \quad (5.54)$$

where d and $\mathbf{Q}_{DOA}$ are given in (5.41) and (5.42), respectively.

5.3 Hybrid TDOA and DOA for Position Estimation

To improve the positioning accuracy, it is always useful to use various observations for position estimation. In the following, hybrid TDOA and DOA estimators will be described for MU position estimation. As we have explained, there are two options to combine TDOA and DOA measurements. The first option is a combination between the results of TDOA and DOA methods after computing the position of MU for each method individually. To show that DAC and hybrid based on weighting estimators will be presented based on TDOA and DOA measurements. The second option is to use TDOA and DOA measurements together to estimate the MU position, and for that the proposed method will be presented in Section 5.4.

5.3.1 Hybrid DAC based on TDOA and DOA Estimator

The principle of hybrid DAC is to combine DAC based on TDOA and DAC based on DOA described in Sections 5.1.2 and 5.2.3, respectively. First, the intersections between lines defined by pairs of DOA measurements are computed. Second, the intersections between hyperbolas defined by sets of TDOA measurements are computed. Finally, all the partial position estimates are combined by giving each estimate a weighting vector as follows [46]

$$\hat{\mathbf{z}}_{ML} = \sum_{k=1}^{K_1} \mathbf{W}_k^{TDOA}\hat{\mathbf{z}}_k^{TDOA} + \sum_{k=1}^{K_2} \mathbf{W}_k^{DOA}\hat{\mathbf{z}}_k^{DOA} \quad (5.55)$$

where

$$\mathbf{W}_k^{TDOA} = [\sum_{k=1}^{K_1}\mathbf{C}_k + \sum_{k=1}^{K_2}\mathbf{C}'_k]^{-1}\mathbf{C}_k \quad (5.56)$$

$$\mathbf{W}_k^{DOA} = [\sum_{k=1}^{K_1}\mathbf{C}_k + \sum_{k=1}^{K_2}\mathbf{C}'_k]^{-1}\mathbf{C}'_k \quad (5.57)$$

where K_1 is the number of sets of TDOA measurements, and K_2 is the number of pairs of DOA measurements. $\mathbf{C}_k$ and $\mathbf{C}'_k$ have been defined in (5.11) and (5.47), respectively.

5.3.2 Weighted TDOA and DOA Estimator

In the previous, the TDOA and DOA observations are used individually to find the position of MU. The XY coordinates of MU position based on the TDOA measurements can be estimated using the presented estimators in Section 5.1. A weighted least square estimator, which will be presented in Section 5.4.1, can also be used. The XY coordinates of MU position based on the DOA measurements can be estimated using the presented estimators in Section 5.2. The final position of MU can be found by using any combination between the presented estimators in Sections 5.1 and 5.2, where each estimated value should be given an appropriate weight calculated from the measurements covariance or variance matrices.

Let us assume the estimated XY coordinates based on TDOA and DOA observations are $\hat{\mathbf{z}}_{TDOA} = [\hat{x}, \hat{y}]^T$ and $\hat{\mathbf{z}}_{DOA} = [\hat{x}, \hat{y}]^T$, respectively. To combine the estimated coordinates, appropriate weights should be used to scale the different accuracy of using TDOA and DOA observations. The variances of TDOA and DOA observations represented in distance form are used. For TDOA, it is obtained by calculating the mean of the diagonal elements of the covariance matrix $\mathbf{Q}_{TDOA}$ converted from its original TDOA form into the corresponding distance form as

$$\sigma^2_{r,TDOA} = \text{avg}(\text{diag}(\mathbf{Q}_{TDOA})). \tag{5.58}$$

For DOA, it is obtained by calculating the mean of the diagonal elements of the variance matrix $\mathbf{Q}_{DOA}$. To convert it to distance form, the unknown distances $\mathbf{d} = \text{diag}(d_1^2, \ldots, d_I^2)$ between the MU and BSs can be calculated from the initial estimation of MU using TDOA measurements, then

$$\sigma^2_{r,DOA} = \text{avg}(\text{diag}(\mathbf{d}\mathbf{Q}_{DOA})). \tag{5.59}$$

Then, the required weights are

$$\kappa_{TDOA} = \frac{\sigma^2_{r,DOA}}{\sigma^2_{r,DOA}+\sigma^2_{r,TDOA}}, \quad \kappa_{DOA} = \frac{\sigma^2_{r,TDOA}}{\sigma^2_{r,DOA}+\sigma^2_{r,TDOA}}. \tag{5.60}$$

Therefore, the final estimate can be obtained:

$$\hat{\mathbf{z}} = \hat{\mathbf{z}}_{TDOA} \cdot \kappa_{TDOA} + \hat{\mathbf{z}}_{DOA} \cdot \kappa_{DOA}. \tag{5.61}$$

5.4 Proposed Hybrid TDOA and DOA Estimator

From the previous, DAC estimator requires that the Fisher information should be sufficiently large, and the noise should be small enough. The methods which are based on the initial guess cannot guarantee the convergence if the initial guess is not close enough to the MU position such as in [119]. Therefore, a direct method will be proposed in two steps. First, the weighted-least square (W-LS) principle is used based on the TDOA measurements only to estimate the initial position of MU, $\hat{\mathbf{z}}_0 = [\hat{x}_0, \hat{y}_0]^T$, since using TDOA observations is more accurate than using DOA observations. Second, a hybrid TDOA and DOA method is used to get the final position estimate.

5.4.1 Initial Position Estimation using Weighted-Least Square Estimator

Based on Fig. 5.1, the distances between MU and BSs can be defined as follows

$$\begin{aligned} d_1^2 &= (x-x_1)^2 + (y-y_1)^2 \\ d_2^2 &= (x-x_2)^2 + (y-y_2)^2 \\ d_3^2 &= (x-x_3)^2 + (y-y_3)^2 \\ &\vdots \\ d_I^2 &= (x-x_I)^2 + (y-y_I)^2 \end{aligned} \tag{5.62}$$

The distances between MU and BSs can also be defined as [6]

$$\begin{aligned} d_2 &= d_{21} + d_1 \\ d_3 &= d_{31} + d_1 \\ &\vdots \\ d_I &= d_{I1} + d_1 \end{aligned} \tag{5.63}$$

By replacing (5.63) into (5.62), we obtained

$$\begin{aligned} d_1^2 &= (x-x_1)^2 + (y-y_1)^2 \\ (d_{21} + d_1)^2 &= (x-x_2)^2 + (y-y_2)^2 \\ (d_{31} + d_1)^2 &= (x-x_3)^2 + (y-y_3)^2 \\ &\vdots \\ (d_{I1} + d_1)^2 &= (x-x_I)^2 + (y-y_I)^2 \end{aligned} \tag{5.64}$$

After some manipulations in (5.64), we can derive a set of linear equations in a function of $\hat{x}$, $\hat{y}$, and $\hat{d}_1$ as follows

$$\begin{aligned} (x_1-x_2)\hat{x} + (y_1-y_2)\hat{y} &= \tfrac{1}{2}(x_1^2 - x_2^2 + y_1^2 - y_2^2 + d_{21}^2 + 2d_{21}\hat{d}_1) \\ (x_1-x_3)\hat{x} + (y_1-y_3)\hat{y} &= \tfrac{1}{2}(x_1^2 - x_3^2 + y_1^2 - y_3^2 + d_{31}^2 + 2d_{31}\hat{d}_1) \\ &\vdots \\ (x_1-x_I)\hat{x} + (y_1-y_I)\hat{y} &= \tfrac{1}{2}(x_1^2 - x_I^2 + y_1^2 - y_I^2 + d_{I1}^2 + 2d_{I1}\hat{d}_1) \end{aligned} \tag{5.65}$$

which can be represented in a matrix form as

$$\mathbf{A}_1.\hat{\mathbf{z}} = \mathbf{b}_1 + \mathbf{b}_2.\hat{d}_1 \tag{5.66}$$

where

$$\mathbf{A}_1 = \begin{bmatrix} x_1 - x_2 & y_1 - y_2 \\ x_1 - x_3 & y_1 - y_3 \\ \vdots & \vdots \\ x_1 - x_I & y_1 - y_I \end{bmatrix}, \tag{5.67}$$

$$\mathbf{b}_1 = \frac{1}{2} \begin{bmatrix} x_1^2 - x_2^2 + y_1^2 - y_2^2 + d_{21}^2 \\ x_1^2 - x_3^2 + y_1^2 - y_3^2 + d_{31}^2 \\ \vdots \\ x_1^2 - x_I^2 + y_1^2 - y_I^2 + d_{I1}^2 \end{bmatrix}, \tag{5.68}$$

5.4. Proposed Hybrid TDOA and DOA Estimator

and
$$\mathbf{b}_2 = [d_{21} \; d_{31} \; \cdots \; d_{I1}]^T. \tag{5.69}$$

Hence, the LS estimate of $\hat{\mathbf{z}}$ is

$$\hat{\mathbf{z}} = \begin{bmatrix} \hat{x} \\ \hat{y} \end{bmatrix} = (\mathbf{A}_1^T \mathbf{A}_1)^{-1} \mathbf{A}_1^T (\mathbf{b}_1 + \mathbf{b}_2.\hat{d}_1). \tag{5.70}$$

To improve the LS estimator presented in (5.70), the covariance matrix of TDOA observations $\mathbf{Q}_{TDOA}$, converted from its TDOA form into the corresponding distance form, should be included as

$$\hat{\mathbf{z}} = (\mathbf{A}_1^T \mathbf{Q}_{TDOA}^{-1} \mathbf{A}_1)^{-1} \mathbf{A}_1^T \mathbf{Q}_{TDOA}^{-1} (\mathbf{b}_1 + \mathbf{b}_2.\hat{d}_1) \tag{5.71}$$

which can be simplified as

$$\hat{\mathbf{z}} = \begin{bmatrix} \hat{x} \\ \hat{y} \end{bmatrix} = \begin{bmatrix} a_1 + c_1.\hat{d}_1 \\ a_2 + c_2.\hat{d}_1 \end{bmatrix}. \tag{5.72}$$

From (5.72), $\hat{x}$ and $\hat{y}$ can be defined in terms of $\hat{d}_1$ as

$$\begin{aligned} \hat{x} &= a_1 + c_1 \hat{d}_1, \\ \hat{y} &= a_2 + c_2 \hat{d}_1. \end{aligned} \tag{5.73}$$

From (5.62), $\hat{d}_1$ has been defined in terms of $\hat{x}$ and $\hat{y}$ as

$$\hat{d}_1^2 = (\hat{x} - x_1)^2 + (\hat{y} - y_1)^2. \tag{5.74}$$

By substitute $\hat{x}$ and $\hat{y}$ of (5.73) into (5.74)

$$\hat{d}_1^2 = (a_1 + c_1 \hat{d}_1 - x_1)^2 + (a_2 + c_2 \hat{d}_1 - y_1)^2. \tag{5.75}$$

After some manipulations, $\hat{d}_1$ can be defined as

$$\hat{d}_1 = \frac{-b \pm \sqrt{b^2 - 4ac}}{2a} \tag{5.76}$$

where

$$\begin{aligned} a &= c_1^2 + c_2^2 - 1, \\ b &= 2c_1(a_1 - x_1) + 2c_2(a_2 - y_1), \\ c &= (a_1 - x_1)^2 + (a_2 - y_1)^2. \end{aligned} \tag{5.77}$$

From (5.76), $\hat{d}_1$ should be the positive value. Therefore, the MU coordinates $(\hat{x}, \hat{y})$ can be estimated directly using (5.76) and (5.72). The proposed W-LS estimator has been introduced in [56] for wireless positioning using only TDOA observations.

5.4.2 Final Position Estimation using Hybrid TDOA and DOA Estimator

As it has been shown in (5.3) and (5.30), the TDOA and DOA mathematical models are non-linear. Therefore, a Taylor series expansion is used to linearize the TDOA and DOA mathematical models [119]:

$$\hat{\mathbf{z}} = \begin{bmatrix} \hat{x} \\ \hat{y} \end{bmatrix} = \begin{bmatrix} \hat{x}_0 \\ \hat{y}_0 \end{bmatrix} + (\mathbf{G}^T \mathbf{Q}^{-1} \mathbf{G})^{-1} \mathbf{G}^T \mathbf{Q}^{-1} \mathbf{h} \qquad (5.78)$$

where $\hat{\mathbf{z}}_0 = [\hat{x}_0, \hat{y}_0]^T$ is the initial estimation of MU position using the previous step, $\mathbf{G}$ is the designed matrix and can be represented as

$$\mathbf{G} = \begin{bmatrix} (x_1 - \hat{x}_0)/\hat{d}_1 - (x_2 - \hat{x}_0)/\hat{d}_2 & (y_1 - \hat{y}_0)/\hat{d}_1 - (y_2 - \hat{y}_0)/\hat{d}_2 \\ (x_1 - \hat{x}_0)/\hat{d}_1 - (x_3 - \hat{x}_0)/\hat{d}_3 & (y_1 - \hat{y}_0)/\hat{d}_1 - (y_3 - \hat{y}_0)/\hat{d}_3 \\ \vdots & \vdots \\ (x_1 - \hat{x}_0)/\hat{d}_1 - (x_I - \hat{x}_0)/\hat{d}_I & (y_1 - \hat{y}_0)/\hat{d}_1 - (y_I - \hat{y}_0)/\hat{d}_I \\ -\sin(\tilde{\theta}_1) & \cos(\tilde{\theta}_1) \\ \vdots & \vdots \\ -\sin(\tilde{\theta}_I) & \cos(\tilde{\theta}_I) \end{bmatrix}, \qquad (5.79)$$

and

$$\mathbf{h} = \begin{bmatrix} d_{21} - \hat{d}_{21} \\ \vdots \\ d_{I1} - \hat{d}_{I1} \\ -(x_1 - \hat{x}_0)\sin(\tilde{\theta}_1) + (y_1 - \hat{y}_0)\cos(\tilde{\theta}_1) \\ \vdots \\ -(x_I - \hat{x}_0)\sin(\tilde{\theta}_I) + (y_I - \hat{y}_0)\cos(\tilde{\theta}_I) \end{bmatrix} \qquad (5.80)$$

where $\hat{d}_i$ ($i = 1, ..., I$) and $\hat{d}_{i1}$ ($i = 2, ..., I$) have been calculated from the initial estimation $\hat{\mathbf{z}}_0 = [\hat{x}_0, \hat{y}_0]^T$ of step one. The combined covariance matrix of both TDOA and DOA measurements is defined

$$\mathbf{Q} = \begin{bmatrix} \mathbf{Q}_{TDOA} & 0 \\ 0 & \mathbf{Q}_{DOA} \end{bmatrix} \qquad (5.81)$$

where $\mathbf{Q}_{DOA}$ is the variance matrix of DOA measurements, which can be represented as

$$\mathbf{Q}_{DOA} = \begin{bmatrix} \sigma_1^2 \hat{d}_1^2 & \cdots & 0 \\ \vdots & \ddots & \vdots \\ 0 & \cdots & \sigma_I^2 \hat{d}_I^2 \end{bmatrix} \qquad (5.82)$$

where σ_i^2 ($i = 1, ..., I$) are the DOA variances. The calculation in (5.78) uses the TDOA and DOA estimates together based on the least square of TDOA and Stansfield of DOA principles. The weighting matrix $\mathbf{Q}$ is used to scale the TDOA and DOA estimates inherently due to the different accuracy of both types, as it will be seen in the measurement chapter.

The proposed estimator calculates the initial position estimate using TDOA observations. Then, it uses the TDOA and DOA observations of all BSs at one time rather than split the observations to small sets such as in DAC estimator, which is sensitive to noisy signals. It does not need to assume that the MU is located in the far field for the initial estimation and then another step to compensate that for the close field such as in Chan estimator. The proposed estimator does not need to assume that $\hat{x}$, $\hat{y}$, and $\hat{d}_1$ are independent in the initial phase and then another step to compensate that such as in Chan estimator, where the initial estimation of $\hat{x}$, $\hat{y}$, and $\hat{d}_1$ is done simultaneously. The proposed estimator can converge using only one iteration. It has been presented in [58].

Finally, we would like to summarize the presented estimators in the previous part of this chapter for MU coordinates estimation based on using TDOA and DOA observations individually or in hybrid. The summary is in Table 5.1. The weighted estimators require the values of matrix Q either for TDOA or DOA observations, where the non-weighted estimators assume that all BSs have the same variance (equal weight). In the measurement chapter, we will show the performance of all previous algorithms for MU coordinates estimation.

5.5 Lower Bounds of TDOA and DOA Estimation Error Variances

The positioning accuracy is limited by the fundamental parameters of the wireless positioning system such as the antenna array order, the number of subcarriers including subcarrier spacing, the SNR, the estimated DOA with respect to the antenna array, and many others. To measure the effect of those parameters on the positioning performance, the time delay and DOA error variances should be calculated. The CRB provides a minimum limit or a lower bound on the variance of any unbiased estimator of an unknown parameter(s) [27], [65]. Therefore, if the time delay and the DOA are unbiased parameters, the estimation error variances are lower bounded by the CRB [120], [121]. The CRB of DOA observations using the ULA has been derived in [122] for narrowband signals without delay spread. In [27], the CRB of joint time delay and DOA estimation has been derived for narrowband signals. Similarly, it has been presented in [120] for a wireless senor network setup. The CRB of DOA estimation of narrowband signals has been extended to the CRB of UWB orthogonal multi-carrier signals in [65]. The main principle is that the Fisher information matrix (FIM) of OFDM signals is the summation of the FIMs obtained from separate observations of the narrowband subcarriers. The CRBs for location estimation accuracy of hybrid TOA / RSS and TDOA/RSS have been derived in [121], [123]. Analysis of positioning systems using wideband antenna arrays, which are not restricted to far-field assumptions, is presented in [124]. To include the effect of BSs locations, the lower bound of positioning error has been presented in [108] using DOA observations and in [125] using TDOA observations. The CRBs of joint time delay and DOA estimation and of time delay estimation with spatial diversity are derived for wideband ULA-OFDM systems, which have not presented in the literature. A single snapshot from the S-CFR is used, which represents the best property of using MP algorithms compared to the other statistical super-resolution algorithms.

If the time delay τ and the DOA θ are unbiased parameters, the error variances are lower bounded by the CRB, denoted by $\sigma^2_{\tau,CRB}$ and $\sigma^2_{\theta,CRB}$. To simplify the CRB derivation, it has been assumed that the received signal comes from the far field with respect to

Table 5.1: List of estimators of MU position estimation.

Estimator Name	Abbreviation
Estimators Based on TDOA Measurements	
Iterative Least Square estimator	ILS
Weighted Iterative Least Square estimator	W-ILS
Least Square estimator (proposed)	LS
Weighted-Least Square estimator (proposed)	W-LS
DAC-Least Square estimator	DAC-LS
Chan estimator	Chan
Estimators Based on DOA Measurements	
Lines Intersection estimator	LI
Least Square estimator	LS
Weighted Least Square estimator	W-LS
Stansfield-Least Square estimator	SLS
Weighted-Stansfield-Least Square estimator	W-SLS
DAC based on lines intersection	DAC-LI
Adapted DAC based on lines intersection	A-DAC-LI
Hybrid Estimators Based on TDOA and DOA Measurements	
Hybrid DAC based on TDOA and DOA estimator	Hybrid-DAC
Weighted TDOA and DOA Estimator	Hybrid-W
Hybrid TDOA and DOA Estimator (proposed)	Hybrid-WLS

the antenna array as shown in Fig. 2.2, and the wireless channel is propagated via a LOS channel, hence, the time delay and DOA can be assumed to be unbiased parameters. In practice, the DME is biased by a NLOS propagation channel as well as a multipath delay spread.

The least square estimate of the S-CFR of the ULA-OFDM signal for the mth antenna and the kth subcarrier is recalled from (4.10) in Section 4.2

$$H_{m,k} = \sum_{l=1}^{L} \beta_{m,l,k} \cdot e^{-j2\pi k \Delta f(\tau_l + m\rho \sin\theta_l/c)} e^{-j2\pi f_c m\rho \sin\theta_l/c} + w_{m,k} \quad (5.83)$$

where $-(N-1)/2 \leq k \leq (N-1)/2$, $m = 0, \ldots, M-1$, the complex channel gain $\beta_{m,l,k} = \alpha_{m,l,k} \cdot e^{-j2\pi f_c \tau_l}$, and $w_{m,k}$ is the AWGN at the mth antenna and the kth subcarrier. It should be noted that time delays and arrival angles are relatively stationary. As it has been stated in the previous to simplify the CRB derivation, let us assume the MU in the far field, and the wireless channel propagated via a single path channel, hence, time delay and DOA are unbiased parameters. From (5.83), the S-CFR will be then

$$H_{m,k} = \beta \cdot e^{-j2\pi k \Delta f(\tau + m\rho \sin\theta/c)} e^{-j2\pi f_c m\rho \sin\theta/c} + w_{m,k} \quad (5.84)$$

where $\beta = \alpha e^{-j2\pi f_c \tau}$.

In the following, the CRB will be derived using multi-antenna multi-carrier systems for DOA, time delay, and TDOA estimation individually. After that the CRB will be derived for joint time delay and DOA estimation using a multipath channel.

5.5.1 DOA Estimation Error Variance

To get the DOA estimation error variance for OFDM signals, let us find it first for a single subcarrier, for example at carrier frequency f_c. By setting $k = 0$ in (5.84), it will be in a vector form (along M antennas)

$$\mathbf{h} = \beta \mathbf{x}(\theta) + \mathbf{w} = \mathbf{u}(\theta) + \mathbf{w} \tag{5.85}$$

where the noise vector $\mathbf{w}$ is zero mean Gaussian with covariance $\mathbf{Q}_n = \sigma_n^2 \mathbf{I}$, the channel gain $\beta = a e^{jb}$ represents two unknowns (its magnitude and phase), and $\mathbf{x}(\theta)$ represents the steering vector of the signal whose direction θ with respect to the ULA, which should be estimated. If the array center is the reference and the number of antenna elements is odd, the steering vector is

$$\mathbf{x}(\theta) = [x^{-(M-1)/2}, x^{-(M-3)/2}, \ldots, x^{-1}, 1, x, \ldots, x^{(M-3)/2}, x^{(M-1)/2}]^T \tag{5.86}$$

where $x = e^{-j2\pi f_c \rho \sin\theta/c}$. The unknown parameters are $\boldsymbol{\eta} = [a, b, \theta]$, which are modeled as deterministic (i.e., fixed) quantities, hence, $E\{\mathbf{h}\} = \mathbf{u} = \beta \mathbf{x}(\theta)$, where $E\{.\}$ represents the statistical expectation.

The error variance of an unbiased estimate of the lth parameter, η_l, is lower bounded by the CRB inequality as follows [124], [122]

$$\sigma_{\eta_l,CRB}^2 \geq \mathbf{F}_{ll}^{-1} \tag{5.87}$$

where $\mathbf{F}_{ll}^{-1}$ is the lth diagonal entry of the inverse of the FIM $\mathbf{F}$ whose (i,j)th is given by

$$\mathbf{F}_{ij} = -E\{\frac{\partial^2}{\partial \eta_i \partial \eta_j}[\ln f_\mathbf{h}(\mathbf{h}/\boldsymbol{\eta})]\} \tag{5.88}$$

where $f_\mathbf{h}(\mathbf{h}/\boldsymbol{\eta})$ is the probability density function (pdf) of the received vector given the parameters $\boldsymbol{\eta}$. In our case, it is Gaussian

$$f_\mathbf{h}(\mathbf{h}/\boldsymbol{\eta}) = C e^{-(\mathbf{h}-\mathbf{u})^H \mathbf{Q}_n^{-1} (\mathbf{h}-\mathbf{u})} \tag{5.89}$$

where C is a normalization constant. To proceed for (5.88), the natural logarithm of (5.89) is

$$g(\boldsymbol{\eta}) = \ln\{f_\mathbf{h}(\mathbf{h}/\boldsymbol{\eta})\} = \ln(C) - \frac{1}{\sigma_n^2}(\mathbf{h}-\mathbf{u})^H(\mathbf{h}-\mathbf{u}) \tag{5.90}$$

which can be simplified to

$$g(\boldsymbol{\eta}) = \ln(C) + \frac{1}{\sigma_n^2}(-\mathbf{h}^H \mathbf{h} + \beta^* \mathbf{x}^H \mathbf{h} + \beta \mathbf{h}^H \mathbf{x} - |\beta|^2 \mathbf{x}^H \mathbf{x}). \tag{5.91}$$

Under the assumption of uncorrelated signals (diagonal $\mathbf{F}$), the result of (5.87) in terms of the interesting parameter θ is

$$\sigma_{\theta,CRB}^2 \geq \{-E[\frac{\partial^2 g(\boldsymbol{\eta})}{\partial \theta^2}]\}^{-1}. \tag{5.92}$$

The first partial derivative of $\mathbf{x}(\theta)$ as a function of θ is

$$\frac{\partial \mathbf{x}(\theta)}{\partial \theta} = \mathbf{x}_1(\theta) = -j2\pi f_c \rho \cos\theta/c[\frac{-(M-1)}{2}x^{-(M-1)/2}, \frac{-(M-3)}{2}x^{-(M-3)/2},$$
$$\ldots, -x^{-1}, 0, x, \ldots, \frac{(M-3)}{2}x^{(M-3)/2}, \frac{(M-1)}{2}x^{(M-1)/2}]^T \tag{5.93}$$

and

$$\frac{\partial \mathbf{x}^H(\theta)}{\partial \theta} = \mathbf{x}_1^H(\theta) = j2\pi f_c \rho \cos\theta/c [\frac{-(M-1)}{2} x^{-(M-1)/2}, \frac{-(M-3)}{2} x^{-(M-3)/2}, \\ \ldots, -x^{-1}, 0, x, \ldots, \frac{(M-3)}{2} x^{(M-3)/2}, \frac{(M-1)}{2} x^{(M-1)/2}]^*. \quad (5.94)$$

The multiplication of the above two equations is

$$\mathbf{x}_1^H(\theta)\mathbf{x}_1(\theta) = (2\pi f_c \rho \cos\theta/c)^2 \sum_{m=\frac{-(M-1)}{2}}^{\frac{(M-1)}{2}} m^2. \quad (5.95)$$

The summation in (5.95) can be simplified to

$$\sum_{m=\frac{-(M-1)}{2}}^{\frac{(M-1)}{2}} m^2 = 2\sum_{m=1}^{\frac{(M-1)}{2}} m^2 = M(M^2-1)/12. \quad (5.96)$$

If the number of antennas is even and the array center is the reference, the steering vector is

$$\mathbf{x}(\theta) = [x^{-(M-2)/2}, x^{-(M-4)/2}, \ldots, x^{-1}, 1, x, \ldots, x^{(M-2)/2}, x^{M/2}]^T. \quad (5.97)$$

Such as in (5.95), the multiplication of the first partial derivative of $\mathbf{x}^H(\theta)$ and $\mathbf{x}(\theta)$ is

$$\mathbf{x}_1^H(\theta)\mathbf{x}_1(\theta) = (2\pi f_c \rho \cos\theta/c)^2 \sum_{m=\frac{-(M-2)}{2}}^{\frac{M}{2}} m^2. \quad (5.98)$$

The summation in (5.98) can be simplified to

$$\sum_{m=\frac{-(M-2)}{2}}^{\frac{M}{2}} m^2 = \frac{M(M-1)(M-2)+3M^2}{12}. \quad (5.99)$$

Finally, if the number of antennas is 2, the steering vector is $\mathbf{x}(\theta) = [1, x]^T$, and the multiplication of the first partial derivative of $\mathbf{x}^H(\theta)$ and $\mathbf{x}(\theta)$ is $(2\pi f_c \rho \cos\theta/c)^2$.

To proceed for (5.92), we should first find the partial derivative of $g(\boldsymbol{\eta})$ as a function of θ, where the first two terms are constants with respect to θ

$$\frac{\partial g}{\partial \theta} = \frac{1}{\sigma_n^2}\{\beta^* \mathbf{x}_1^H \mathbf{h} + \beta \mathbf{h}^H \mathbf{x}_1 - a^2(\mathbf{x}_1^H \mathbf{x} + \mathbf{x}^H \mathbf{x}_1)\}. \quad (5.100)$$

The second partial derivative of $g(\boldsymbol{\eta})$ with respect to θ is

$$\frac{\partial^2 g}{\partial \theta^2} = \frac{1}{\sigma_n^2}\{\beta^* \mathbf{x}_2^H \mathbf{h} + \beta \mathbf{h}^H \mathbf{x}_2 - a^2(\mathbf{x}_2^H \mathbf{x} + \mathbf{x}_1^H \mathbf{x}_1 + \mathbf{x}_1^H \mathbf{x}_1 + \mathbf{x}^H \mathbf{x}_2)\}, \quad (5.101)$$

$$\Rightarrow E\{\frac{\partial^2 g}{\partial \theta^2}\} = \frac{1}{\sigma_n^2}\{|\beta|^2 \mathbf{x}_2^H \mathbf{x} + |\beta|^2 \mathbf{x}^H \mathbf{x}_2 - a^2(\mathbf{x}_2^H \mathbf{x} + \mathbf{x}_1^H \mathbf{x}_1 + \mathbf{x}_1^H \mathbf{x}_1 + \mathbf{x}^H \mathbf{x}_2)\}, \quad (5.102)$$

$$\Rightarrow E[\frac{\partial^2 g}{\partial \theta^2}] = \frac{-2a^2}{\sigma_n^2}\mathbf{x}_1^H \mathbf{x}_1. \quad (5.103)$$

5.5. Lower Bounds of TDOA and DOA Estimation Error Variances

Substituting the term $\mathbf{x}_1^H \mathbf{x}_1$ from the above equations based on the order of antenna array into (5.103) gives

$$\Rightarrow E\{\frac{\partial^2 g}{\partial \theta^2}\} = \frac{-a^2}{6\sigma_n^2}(2\pi f_c \rho \cos\theta/c)^2 B \tag{5.104}$$

where the value of B depends on the order of antenna array which is

$$B = \begin{cases} M(M^2 - 1) & \text{if } M = \text{odd } \& M \neq 1 \\ M(M-1)(M-2) + 3M^2 & \text{if } M = \text{even } \& M \neq 2 \\ 12 & \text{if } M = 2 \end{cases} \tag{5.105}$$

From (5.92) and (5.104), the CRB variance of DOA estimation is

$$\sigma_{\theta,CRB}^2 \geq \frac{6}{(2\pi f_c \rho \cos\theta/c)^2 \gamma_{sc} B} \tag{5.106}$$

where $\gamma_{sc} = a^2/\sigma_n^2$ is the SNR per subcarrier. It should be noted the following from the above relation:

- The factor $(2\pi f_c \rho \cos\theta/c)$ represents the effect of the orientation of the ULA (array axis) with respect to the impinging wave direction. If the array axis is perpendicular to the impinging wave ($\theta = 0°$), the CRB of DOA estimation is the minimum value, but if the array axis is aligned to the same direction of the impinging wave ($\theta = \pm\pi/2$), the CRB tends to infinity.

- If the SNR γ_{sc} increases, the error variance decreases.

- Increasing the number of antenna elements in the ULA (M) improves the accuracy of the DOA estimation.

Similar to the principle in [65], the FIM of OFDM signals is the summation of the N FIMs obtained from separate observations of the N narrowband subcarriers, where N is the length of the estimated CFR of the OFDM signal. Hence, the FIM of OFDM signal is

$$\mathbf{F}_{OFDM} = \sum_{k=0}^{N-1} \mathbf{F}_{sc,k} \tag{5.107}$$

where the subscript $_{sc}$ denotes the FIM $\mathbf{F}_k$ of a single subcarrier. The CRB of DOA estimation for OFDM signals is then given by

$$\sigma_{\theta,CRB,OFDM}^2 \geq \{\mathbf{F}_{OFDM}\}^{-1}. \tag{5.108}$$

For simplicity, let us mention to the minimum frequency among subcarriers by f_0 and $f_k = f_0 + k\Delta f$, where $k = 0, \ldots, N-1$. As a consequence, the CRB variance of DOA estimation for OFDM signals can be evaluated easily from (5.106) and (5.107) as

$$\sigma_{\theta,CRB,OFDM}^2 \geq \frac{6}{(2\pi \rho \cos\theta/c)^2 B \gamma_{sc} \sum_{k=0}^{N-1} f_k^2} \tag{5.109}$$

where the value of B is given in (5.105).

From (5.109), the main system parameters affecting on the CRB based on the DOA measurements are the SNR per subcarrier γ_{sc}, the order of antenna array M, the system

BW represented by the number of subcarriers N, and the DOA of the received signal θ. To show the effect of the above parameters, assume the discussed parameter is variable and the remaining parameters are constants with the following values: the number of antennas M is 4, the system BW is 20 MHz, the DOA is 5 degrees, and γ_{sc} is 20 dB. The carrier frequency f_c is 5.25 GHz, and the spacing between the adjacent antenna elements ρ is the half wave length. To show the effect of SNR, Fig. 5.2 (left) shows the standard deviation (STD) of DOA in degrees with respect to SNR per subcarrier. To show the effect of DOA of the received signal, Fig. 5.2 (right) shows the STD of DOA in degrees with respect to the DOA of the received signal. To show the effect of a number of antennas in the ULA, Fig. 5.3 (left) shows the STD of DOA in degrees with respect to the number of antennas. To show the effect of a system BW, Fig. 5.3 (right) shows the STD of DOA in degrees with respect to the system BW. It has been preferred in the previous to plot the STD of DOA estimation rather than the DME, because the DME depends on the distance between the MU and the BS in case of using DOA measurements. From Fig. 5.2 (right), the MU position with respect to BS could be in the critical region ($\theta \approx \pm\pi/2$), therefore, the number of BSs in the positioning system should be enough for reliable accuracy. In addition, from Fig. 5.3, instead of increasing the antenna array order (system complexity) to increase the accuracy of DOA estimation, the system BW can be increased to get a high accuracy.

5.5.2 TDOA and Time Delay Estimation Error Variance

To derive the CRB error variance of time delay estimation for OFDM signals, let us find it first for the reference antenna ($m = 0$). Substituting m into (5.84) yields

$$H_k = \beta e^{-j2\pi k \Delta f \tau} + w_k \tag{5.110}$$

which can be written in a vector form:

$$\mathbf{h} = \beta \mathbf{z}(\tau) + \mathbf{w} = \mathbf{v}(\tau) + \mathbf{w} \tag{5.111}$$

where $\mathbf{z}(\tau)$ represents the steering vector of the signal whose time delay τ, which should be estimated. If the carrier frequency f_c is the reference and the number of OFDM sensors (pilots) is odd, which represents the CFR length, the steering vector is

$$\mathbf{z}(\tau) = [z^{-(N-1)/2}, z^{-(N-3)/2}, \ldots, z^{-1}, 1, z, \ldots, z^{(N-3)/2}, z^{(N-1)/2}]^T \tag{5.112}$$

where $z = e^{-j2\pi\Delta f \tau}$. The number of OFDM pilots per one OFDM symbol is usually even, however, an interpolation is made for the dc subcarriers as it has been explained in Section 3.3.2. The unknown parameters are $\mathbf{t} = [a, b, \tau]$, which are modeled as deterministic parameters, hence, $E[\mathbf{h}] = \mathbf{v} = \beta \mathbf{z}(\tau)$.

Such as the previous, under the assumption of uncorrelated signals (diagonal $\mathbf{F}$), the CRB of time delay estimation is

$$\sigma^2_{\tau,CRB,OFDM} \geq \{-E[\frac{\partial^2 g(\mathbf{t})}{\partial \tau^2}]\}^{-1} \tag{5.113}$$

where $g(\mathbf{t})$ is the natural logarithm of (5.89), which is a function of time delay τ here, as

$$g(\mathbf{t}) = ln(C) + \frac{1}{\sigma_n^2}(-\mathbf{h}^H\mathbf{h} + \beta^*\mathbf{z}^H\mathbf{h} + \beta\mathbf{h}^H\mathbf{z} - |\beta|^2\mathbf{z}^H\mathbf{z}). \tag{5.114}$$

5.5. Lower Bounds of TDOA and DOA Estimation Error Variances

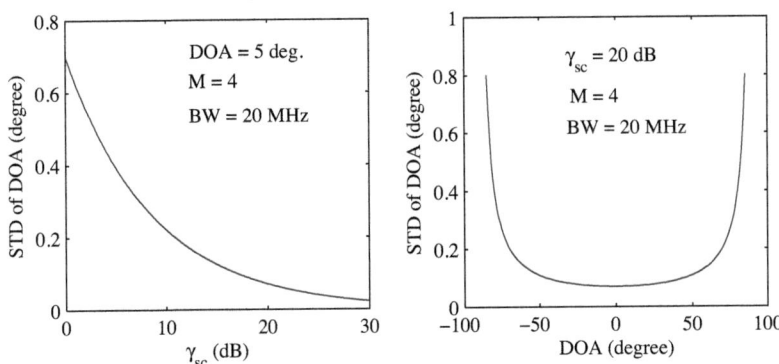

Figure 5.2: STD of DOA estimates versus γ_{sc} (left) and DOA of the received signal (right).

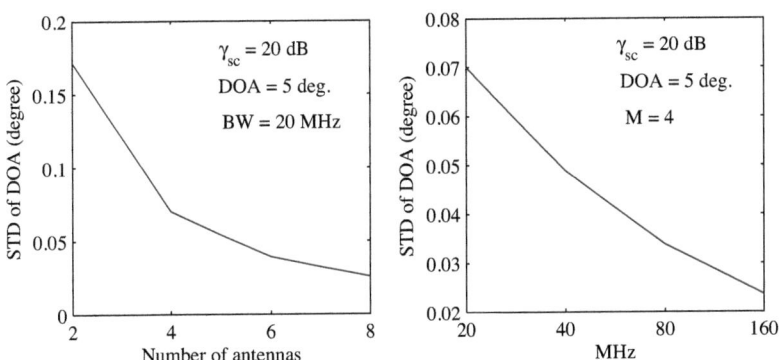

Figure 5.3: STD of DOA estimates versus number of antennas (left) and system BW (right).

Using the previous procedure of DOA error variance gives a solution similar to (5.103)

$$\Rightarrow E\left\{\frac{\partial^2 g(\mathbf{t})}{\partial \tau^2}\right\} = \frac{-2a^2}{\sigma_n^2}\mathbf{z}_1^H \mathbf{z}_1. \tag{5.115}$$

Such as in (5.95), the multiplication of the first partial derivative of the time delay steering vectors $\mathbf{z}^H(\tau)$ and $\mathbf{z}(\tau)$ is

$$\mathbf{z}_1^H(\tau)\mathbf{z}_1(\tau) = (2\pi\Delta f)^2 N(N^2 - 1)/12. \tag{5.116}$$

From (5.113), (5.115), and (5.116), the CRB of time delay estimation for OFDM signals is

$$\sigma^2_{\tau,CRB,OFDM} \geq \frac{6}{(2\pi\Delta f)^2 N(N^2 - 1)\gamma_{sc}} \tag{5.117}$$

where γ_{sc} is the SNR per subcarrier. It should be noted the following from the above relation:

- Increasing the number of subcarriers including the subcarrier spacing decreases the CRB error variance of time delay estimation, which improves the accuracy of positioning.

- If the SNR γ_{sc} increases, the error variance decreases.

Let us now extend the CRB of time delay estimation using OFDM signals to include the effect of multiple snapshot principle based on the spatial diversity. The S-CFR samples using M antennas and N subcarriers presented in (5.84) can be written in a vector form along the subcarriers for the mth antenna as follows

$$\mathbf{h}_m = \beta \mathbf{z}_m(\tau) + \mathbf{w}_m \tag{5.118}$$

where the time delay steering vector is

$$\mathbf{z}_m(\tau) = e^{-j2\pi f_c m\rho \sin\theta/c}[z_m^{-(N-1)/2}, z_m^{-(N-3)/2}, \ldots, z_m^{-1}, 1, z_m, \ldots, z_m^{(N-3)/2}, z_m^{(N-1)/2}]^T \tag{5.119}$$

where $z_m = e^{-j2\pi\Delta f(\tau + m\rho \sin\theta/c)}$. The first partial derivative of $\mathbf{z}_m(\tau)$ as a function of τ is

$$\frac{\partial \mathbf{z}_m(\tau)}{\partial \tau} = \mathbf{z}_{m,1}(\tau) = -j2\pi \Delta f e^{-j2\pi f_c m\rho \sin\theta/c}[\frac{-(N-1)}{2}z_m^{-(N-1)/2}, \frac{-(N-3)}{2}z_m^{-(N-3)/2},$$
$$\ldots, -z_m^{-1}, 0, z_m, \ldots, \frac{(N-3)}{2}z_m^{(N-3)/2}, \frac{(N-1)}{2}z_m^{(N-1)/2}]^T. \tag{5.120}$$

The multiplication of the first partial derivative of the time delay steering vectors is

$$\mathbf{z}_{m,1}^H(\tau)\mathbf{z}_{m,1}(\tau) = (2\pi\Delta f)^2 N(N^2 - 1)/12. \tag{5.121}$$

And then proceed as before to obtain

$$E\left\{\frac{\partial^2 g(\mathbf{t})}{\partial \tau^2}\right\} = \frac{-2a^2}{\sigma_n^2}\mathbf{z}_{m,1}^H \mathbf{z}_{m,1} = \frac{-a^2}{6\sigma_n^2}(2\pi\Delta f)^2 N(N^2 - 1). \tag{5.122}$$

5.5. Lower Bounds of TDOA and DOA Estimation Error Variances

From the result in (5.122), we can assume without approve that the FIM of the OFDM signal with spatial diversity is the summation of the M FIMs obtained from separate observations of M antennas

$$\mathbf{F}_{OFDM,SD} = \sum_{m=0}^{M-1} \mathbf{F}_{OFDM,m} \qquad (5.123)$$

where the subscript $_{SD}$ stands for the spatial diversity. From (5.123), the FIM of the OFDM signal using antenna array with M antennas is

$$\mathbf{F}_{OFDM,SD} = -E\{\frac{\partial^2 g(\mathbf{t})}{\partial \tau^2}\}.M = \frac{a^2}{6\sigma_n^2}(2\pi\Delta f)^2 N(N^2-1)M. \qquad (5.124)$$

Finally, the CRB of time delay estimation for OFDM signals with spatial diversity is

$$\sigma^2_{\tau,CRB,OFDM,SD} \geq \frac{6}{(2\pi\Delta f)^2 N(N^2-1)M\gamma_{sc}}. \qquad (5.125)$$

It is worth mentioning from (5.125) that the accuracy of time delay estimation does not affected by the orientation of antenna arrays with respect to the MU in case of using more than one antenna for spatial diversity.

The DME variance can then be obtained by multiply the time delay (TOA) error variance with c^2 as, $\sigma^2_{r,TOA} = c^2\sigma^2_\tau$. The DME variance using TDOA measurements has been computed in [125] for cellular positioning as

$$\sigma^2_{r,TDOA} = 2\sigma^2_{r,TOA}. \qquad (5.126)$$

However, the positioning error variance using TDOA measurements depends on the BSs locations with respect to the reference BS. Assume the number of BSs in the localization system is I, the CRB on the positioning error variance using TDOA measurements has been calculated in [125] as

$$\sigma^2_{r,TDOA} = \frac{4\sigma^2_{r,TOA}\sum_{i=2}^{I}\{1-\cos(\theta_i-\theta_1)\}}{\sum_{i=2}^{I}\sum_{\substack{j=2 \\ j\neq i}}^{I}\{\sin(\theta_j-\theta_i)+\sin(\theta_i-\theta_1)+\sin(\theta_1-\theta_j)\}^2} \qquad (5.127)$$

where $\theta_i, \{i=1,\ldots,I\}$, is the angle from BS number i to the MU. In the above relation, a LOS single path has been assumed to all BSs. In practice, the time delay error variance of each BS depends on its channel profile.

From (5.125) and (5.127), the CRB of the positioning error variance using TDOA measurements depends on the SNR per subcarrier γ_{sc}, the order of antenna array M, the system BW represented by the number of subcarriers N and the subcarrier spacing Δf, and the BSs locations. Let us assume we have four BSs, and the DOAs are 46.5, 135.3, -158.3, and -21.2 degrees. To show the effect of the above parameters, assume the discussed parameter is variable, and the remaining parameters are constants with the following values, the system BW is 20 MHz, the number of antennas M is 4, and γ_{sc} is 20 dB. The carrier frequency f_c is 5.25 GHz, and the spacing between the adjacent antenna elements is the half wave length. To show the effect of SNR, Fig. 5.4 shows the STD of time delay and positioning using TDOA error variances in nanoseconds and in meters, respectively, with respect to SNR per subcarrier γ_{sc}. To show the effect of system bandwidth, Fig. 5.5 shows the STD of time delay and positioning using TDOA error variances with respect to the system BW. To show the effect of a number of antennas, Fig. 5.6 shows the STD of time delay and positioning using TDOA error variances with respect to the number of antennas. From Fig. 5.5 and 5.6, it can conclude that increasing system BW is more useful than increasing number of antennas for spatial diversity.

5.5.3 Joint Time Delay and DOA Estimation Error Variance

In the previous, some assumptions have been assumed to derive the CRB of time delay and DOA error variances individually to let us present the effect of the main system parameters clearly. In this section, the derivation of the CRB of joint time delay and DOA estimation using a single snapshot from the S-CFR will be presented for the wireless multipath channel. The noiseless S-CFR has been recalled from (4.10)

$$H_{m,k} = \sum_{l=1}^{L} \beta_{m,l,k} x_l^m z_l^k \qquad (5.128)$$

where $x(\theta_l) = e^{-j2\pi f_c \rho \sin \theta_l / c}$ and $z(\tau_l) = e^{-j2\pi \Delta f \tau_l}$. By assuming that the MU is in the far field, the channel complex gain of path l in (5.128) is $\alpha_l \approx \alpha_{m,l,k}$, hence, $\beta_l = \alpha_l e^{-j2\pi f_c \tau_l}$. The noisy channel estimate is written in a matrix form:

$$\hat{\mathbf{H}} = \mathbf{H} + \mathbf{W} \qquad (5.129)$$

where $\mathbf{W}$ is the additive noise. The matrix $\mathbf{H}$ can be factorized as in (4.15) to

$$\mathbf{H} = [\mathbf{x}(\theta_1), \ldots, \mathbf{x}(\theta_L)] \begin{bmatrix} \beta_1 & & 0 \\ & \ddots & \\ 0 & & \beta_L \end{bmatrix} [\mathbf{z}(\tau_1), \ldots, \mathbf{z}(\tau_L)]^T \qquad (5.130)$$

where the steering vector $\mathbf{x}(\theta_l)$ of θ_l is determined from the antenna geometry:

$$\mathbf{x}(\theta_l) = [1, x_l, \ldots, x_l^{M-1}]^T \qquad (5.131)$$

and the steering vector $\mathbf{z}(\tau_l)$ of τ_l is determined from the OFDM sensors (pilots):

$$\mathbf{z}(\tau_l) = [z_l^{-(N-1)/2}, z_l^{-(N-3)/2}, \ldots, z_l^{-1}, 1, z_l, \ldots, z_l^{(N-3)/2}, z_l^{(N-1)/2}]^T \qquad (5.132)$$

Eq. (5.130) can be simply written:

$$\mathbf{H} = \mathbf{X}(\boldsymbol{\theta}) \mathbf{A}_d \mathbf{Z}^T(\boldsymbol{\tau}) \qquad (5.133)$$

where $\mathbf{A}_d = \text{diag}\{\mathbf{A}_0\}$, $\mathbf{A}_0 = [\beta_1, \ldots, \beta_L]^T$, $\boldsymbol{\theta} = [\theta_1, \ldots, \theta_L]$, and $\boldsymbol{\tau} = [\tau_1, \ldots, \tau_L]$. By using the general relation $vec[\mathbf{C}\text{diag}(\mathbf{a})\mathbf{B}] = (\mathbf{B}^T \odot \mathbf{C})\mathbf{a}$ [27] (the vector operation of a matrix $\mathbf{H}$ is obtained by stacking each column of the matrix $\mathbf{H}$ one under another), we obtain

$$vec[\mathbf{H}] = [\mathbf{Z}(\boldsymbol{\tau}) \odot \mathbf{X}(\boldsymbol{\theta})]\mathbf{A}_0 = \mathbf{U}(\boldsymbol{\theta}, \boldsymbol{\tau})\mathbf{A}_0 \qquad (5.134)$$

which is a column-wise Kronecker product as in (B.2) of appendix B

$$\mathbf{Z} \odot \mathbf{X} = [\mathbf{z}_1 \otimes \mathbf{x}_1, \mathbf{z}_2 \otimes \mathbf{x}_2, \ldots] \qquad (5.135)$$

5.5. Lower Bounds of TDOA and DOA Estimation Error Variances

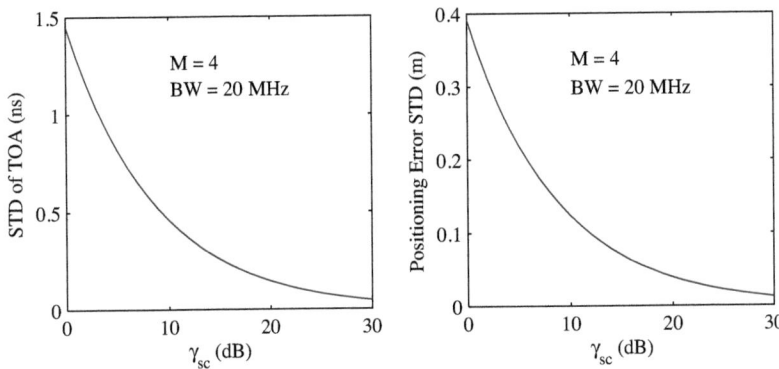

Figure 5.4: STD of time delay and positioning using TDOA error variances versus γ_{sc}.

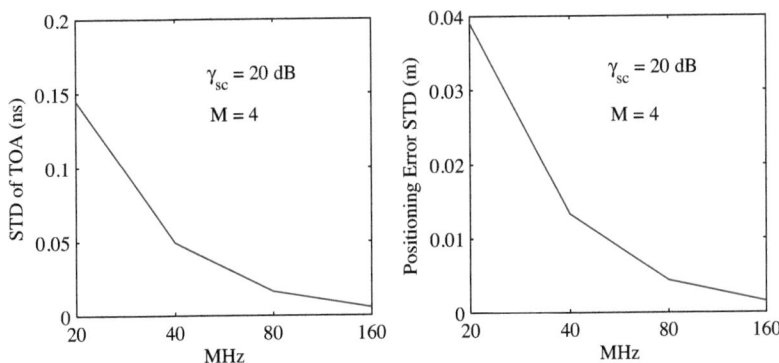

Figure 5.5: STD of time delay and positioning using TDOA error variances versus system BW.

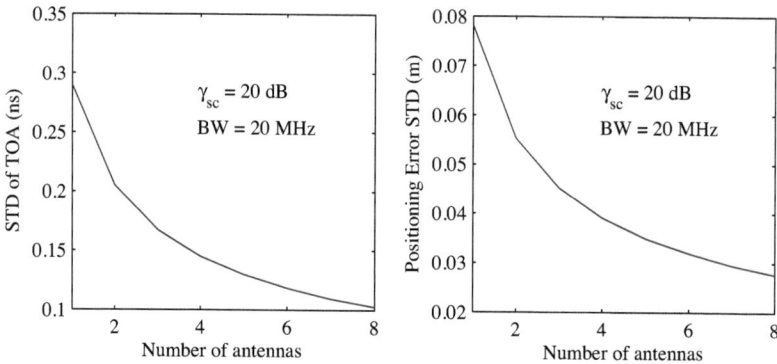

Figure 5.6: STD of time delay and positioning using TDOA error variances versus M.

where $\odot$ and $\otimes$ are the Khatri-Rao and the Kronecker products as in appendix B. The $MN \times L$ matrix $\mathbf{U}(\boldsymbol{\theta}, \boldsymbol{\tau})$ represents the space-frequency response matrix with L paths. For a single path, the space-frequency response vector $\mathbf{u}(\theta, \tau)$ is

$$\mathbf{u}(\theta, \tau) = \mathbf{z}(\tau) \otimes \mathbf{x}(\theta). \tag{5.136}$$

Consequently, the space-frequency response matrix for L paths is defined as in (5.135) as

$$\mathbf{U}(\boldsymbol{\theta}, \boldsymbol{\tau}) = \mathbf{Z}(\boldsymbol{\tau}) \odot \mathbf{X}(\boldsymbol{\theta}) = [\mathbf{u}(\theta_1, \tau_1), \ldots, \mathbf{u}(\theta_L, \tau_L)]. \tag{5.137}$$

The general form of (5.134) including noise is

$$\mathbf{y} = vec[\hat{\mathbf{H}}] = \mathbf{U}(\boldsymbol{\theta}, \boldsymbol{\tau})\mathbf{A}_0 + \mathbf{w} \tag{5.138}$$

where $\mathbf{w} = vec[\hat{\mathbf{W}}]$ is a complex, stationary, and zero mean Gaussian process. It is uncorrelated from path to path, which means $E\{\mathbf{w}\mathbf{w}^H\} = \sigma_n^2 \mathbf{I}$ and $E\{\mathbf{w}\mathbf{w}^T\} = 0$.

The CRB depends on the modeling of path fading, which can be modeled as random variables with a known distribution or as unknown deterministic parameters [27]. In this work, the CRB of deterministic parameters is derived, which can be written in a vector form as, $\boldsymbol{\Omega} = [\sigma_n^2, \mathbf{A}_{Re}^T, \mathbf{A}_{Im}^T, \boldsymbol{\theta}, \boldsymbol{\tau}]$, where $\mathbf{A}_{Re} = \text{Re}[\mathbf{A}_0]$, and $\mathbf{A}_{Im} = \text{Im}[\mathbf{A}_0]$. The likelihood function of the channel estimates $\mathbf{y}$ (independent and identical Gaussian distribution) is

$$f_{\mathbf{y}}(\mathbf{y}/\boldsymbol{\Omega}) = \frac{1}{(2\pi)^{MN}(\sigma_n^2/2)^{MN}}.exp\{-\frac{1}{\sigma_n^2}[\mathbf{y} - \mathbf{U}\mathbf{A}_0]^H[\mathbf{y} - \mathbf{U}\mathbf{A}_0]\}. \tag{5.139}$$

The log-likelihood function of (5.139) is

$$ln\{f_\mathbf{y}\} = C - MN ln\sigma_n^2 - \frac{1}{\sigma_n^2}[\mathbf{y} - \mathbf{U}\mathbf{A}_0]^H[\mathbf{y} - \mathbf{U}\mathbf{A}_0] \tag{5.140}$$

where C is a constant parameter, and can be ignored. Hence, (5.140) can be redefined as

$$g(\boldsymbol{\Omega}) = -MN ln\sigma_n^2 - \frac{1}{\sigma_n^2}[\mathbf{y} - \mathbf{U}\mathbf{A}_0]^H[\mathbf{y} - \mathbf{U}\mathbf{A}_0] \tag{5.141}$$

5.5. Lower Bounds of TDOA and DOA Estimation Error Variances

which can be simplified to

$$g(\Omega) = -MN\ln\sigma_n^2 + \frac{1}{\sigma_n^2}\{-\mathbf{y}^H\mathbf{y} + \mathbf{y}^H\mathbf{U}\mathbf{A}_0 + \mathbf{A}_0^H\mathbf{U}^H\mathbf{y} - \mathbf{A}_0^H\mathbf{U}^H\mathbf{U}\mathbf{A}_0\}. \quad (5.142)$$

To calculate the FIM, the above equation should be derived with respect to σ_n^2, $\mathbf{A}_{Re}$, $\mathbf{A}_{Im}$, θ, and τ as follows:

$$\frac{\partial g}{\partial \sigma_n^2} = \frac{-MN}{\sigma_n^2} + \frac{1}{\sigma_n^4}\mathbf{w}^H\mathbf{w} \quad (5.143)$$

$$\frac{\partial g}{\partial \mathbf{A}_{Re}} = \frac{1}{\sigma_n^2}[\mathbf{y}^H\mathbf{U} + \mathbf{U}^H\mathbf{y} - \mathbf{U}^H\mathbf{U}\mathbf{A}_0 - \mathbf{A}_0^H\mathbf{U}^H\mathbf{U}]$$

$$\Rightarrow \frac{\partial g}{\partial \mathbf{A}_{Re}} = \frac{1}{\sigma_n^2}[\mathbf{w}^H\mathbf{U} + \mathbf{U}^H\mathbf{w}] = \frac{1}{\sigma_n^2}[(\mathbf{U}^H\mathbf{w})^H + \mathbf{U}^H\mathbf{w}]$$

$$\Rightarrow \frac{\partial g}{\partial \mathbf{A}_{Re}} = \frac{2}{\sigma_n^2}\mathrm{Re}[\mathbf{U}^H\mathbf{w}] \quad (5.144)$$

$$\frac{\partial g}{\partial \mathbf{A}_{Im}} = \frac{1}{\sigma_n^2}[j\mathbf{y}^H\mathbf{U} - j\mathbf{U}^H\mathbf{y} + j\mathbf{U}^H\mathbf{U}\mathbf{A}_0 - j\mathbf{A}_0^H\mathbf{U}^H\mathbf{U}]$$

$$\Rightarrow \frac{\partial g}{\partial \mathbf{A}_{Im}} = \frac{j}{\sigma_n^2}[(\mathbf{U}^H\mathbf{w})^H - \mathbf{U}^H\mathbf{w}]$$

$$\Rightarrow \frac{\partial g}{\partial \mathbf{A}_{Im}} = \frac{2}{\sigma_n^2}\mathrm{Im}[\mathbf{U}^H\mathbf{w}] \quad (5.145)$$

$$\frac{\partial g}{\partial \theta_l} = \frac{1}{\sigma_n^2}\frac{\partial}{\partial \theta_l}[\mathbf{y}^H\mathbf{U}\beta_l + \beta_l^*\mathbf{U}^H\mathbf{y} - \beta_l^*\mathbf{U}^H\mathbf{U}\beta_l]$$

$$\Rightarrow \frac{\partial g}{\partial \theta_l} = \frac{2}{\sigma_n^2}\mathrm{Re}[\beta_l^*\mathbf{d}_{\theta_l}^H\mathbf{w}]; l = 1, \ldots, L \quad (5.146)$$

where $\mathbf{d}_{\theta_l}$ is the derivative with respect to θ_l of the lth column of $\mathbf{U}$, which is $\mathbf{d}_{\theta_l} = \mathbf{z}(\tau_l) \otimes \partial\mathbf{x}(\theta_l)/\partial\theta_l$ (see (5.136)). Similarly, the derivative of $g(\Omega)$ with respect to τ_l is

$$\Rightarrow \frac{\partial g}{\partial \tau_l} = \frac{2}{\sigma_n^2}\mathrm{Re}[\beta_l^*\mathbf{d}_{\tau_l}^H\mathbf{w}]; l = 1, \ldots, L \quad (5.147)$$

where $\mathbf{d}_{\tau_l} = \partial\mathbf{z}(\tau_l)/\partial\tau_l \otimes \mathbf{x}(\theta_l)$. From (5.137), the previous two equations (5.146) and (5.147) can be written compactly:

$$\Rightarrow \frac{\partial g}{\partial \theta} = \frac{2}{\sigma_n^2}\mathrm{Re}[\mathbf{A}_d^H\mathbf{D}_\theta^H\mathbf{w}] \quad (5.148)$$

$$\Rightarrow \frac{\partial g}{\partial \tau} = \frac{2}{\sigma_n^2}\mathrm{Re}[\mathbf{A}_d^H\mathbf{D}_\tau^H\mathbf{w}] \quad (5.149)$$

where $\mathbf{D}_\theta = \partial\mathbf{U}/\partial\theta = [\mathbf{d}_{\theta_1}, \ldots, \mathbf{d}_{\theta_L}] = \mathbf{Z} \odot \mathbf{X}_1$, and $\mathbf{D}_\tau = \partial\mathbf{U}/\partial\tau = [\mathbf{d}_{\tau_1}, \ldots, \mathbf{d}_{\tau_L}] = \mathbf{Z}_1 \odot \mathbf{X}$, where the index 1 means the first derivative with respect to the appropriate parameters. Similarly to the procedure of [27], the first derivative of the log-likelihood function $g(\Omega)$ with respect to the interesting parameters $\eta = [\theta, \tau]^T$ can be written as

$$\frac{\partial g}{\partial \eta} = \frac{2}{\sigma_n^2}\mathrm{Re}[\beta_1^*\mathbf{d}_{\theta_1}^H\mathbf{w}, \ldots, \beta_L^*\mathbf{d}_{\theta_L}^H\mathbf{w}, \beta_1^*\mathbf{d}_{\tau_1}^H\mathbf{w}, \ldots, \beta_L^*\mathbf{d}_{\tau_L}^H\mathbf{w}] \quad (5.150)$$

which can be written in a matrix form:

$$\frac{\partial g}{\partial \eta} = \frac{2}{\sigma_n^2}\text{Re}[\mathbb{A}^H \mathbf{D}^H \mathbf{w}] \qquad (5.151)$$

where $\mathbf{D} = [\mathbf{D}_\theta \mathbf{D}_\tau]$ and $\mathbb{A} = \mathbf{I}_2 \otimes \mathbf{A}_d$. To simplify for the final result, the following relations will be used [122]:

$$\text{Re}(\mathbf{B})\text{Re}(\mathbf{C}^T) = \tfrac{1}{2}\{\text{Re}(\mathbf{BC}^T) + \text{Re}(\mathbf{BC}^H)\}, \qquad (5.152)$$
$$\text{Im}(\mathbf{B})\text{Im}(\mathbf{C}^T) = -\tfrac{1}{2}\{\text{Re}(\mathbf{BC}^T) - \text{Re}(\mathbf{BC}^H)\}, \qquad (5.153)$$
$$\text{Re}(\mathbf{B})\text{Im}(\mathbf{C}^T) = \tfrac{1}{2}\{\text{Im}(\mathbf{BC}^T) - \text{Im}(\mathbf{BC}^H)\}. \qquad (5.154)$$

By using the results proven in [122], it has been found that $\partial g/\partial \sigma_n^2$ is not correlated with the other derivatives in the above. Then, by using (5.152) to (5.154) and the fact that $E\{\mathbf{ww}^T\} = 0$ and $E\{\mathbf{ww}^H\} = \sigma_n^2 \mathbf{I}$, we obtain

$$E\{(\frac{\partial g}{\partial \sigma_n^2})^2\} = \frac{MN}{\sigma_n^4} \qquad (5.155)$$

$$E\{(\frac{\partial g}{\partial \mathbf{A}_{Re}})(\frac{\partial g}{\partial \mathbf{A}_{Re}})^T\} = \frac{2}{\sigma_n^4} E\{\text{Re}[\mathbf{U}^H \mathbf{ww}^T \mathbf{U}^*] + \text{Re}[\mathbf{U}^H \mathbf{ww}^H \mathbf{U}]\}$$
$$\Rightarrow E\{(\frac{\partial g}{\partial \mathbf{A}_{Re}})(\frac{\partial g}{\partial \mathbf{A}_{Re}})^T\} = \frac{2}{\sigma_n^2}\text{Re}[\mathbf{U}^H \mathbf{U}] \qquad (5.156)$$

$$E\{(\frac{\partial g}{\partial \mathbf{A}_{Re}})(\frac{\partial g}{\partial \mathbf{A}_{Im}})^T\} = \frac{2}{\sigma_n^4} E\{\text{Im}[\mathbf{U}^H \mathbf{ww}^T \mathbf{U}^*] - \text{Im}[\mathbf{U}^H \mathbf{ww}^H \mathbf{U}]\}$$
$$\Rightarrow E\{(\frac{\partial g}{\partial \mathbf{A}_{Re}})(\frac{\partial g}{\partial \mathbf{A}_{Im}})^T\} = \frac{-2}{\sigma_n^2}\text{Im}[\mathbf{U}^H \mathbf{U}] \qquad (5.157)$$

$$E\{(\frac{\partial g}{\partial \mathbf{A}_{Im}})(\frac{\partial g}{\partial \mathbf{A}_{Im}})^T\} = \frac{-2}{\sigma_n^4} E\{\text{Re}[\mathbf{U}^H \mathbf{ww}^T \mathbf{U}^*] - \text{Re}[\mathbf{U}^H \mathbf{ww}^H \mathbf{U}]\}$$
$$\Rightarrow E\{(\frac{\partial g}{\partial \mathbf{A}_{Im}})(\frac{\partial g}{\partial \mathbf{A}_{Im}})^T\} = \frac{2}{\sigma_n^2}\text{Re}[\mathbf{U}^H \mathbf{U}] \qquad (5.158)$$

$$E\{(\frac{\partial g}{\partial \mathbf{A}_{Re}})(\frac{\partial g}{\partial \eta})^T\} = \frac{2}{\sigma_n^4} E\{\text{Re}[\mathbf{U}^H \mathbf{ww}^T \mathbf{D}^* \mathbb{A}^*] + \text{Re}[\mathbf{U}^H \mathbf{ww}^H \mathbf{D} \mathbb{A}]\}$$
$$\Rightarrow E\{(\frac{\partial g}{\partial \mathbf{A}_{Re}})(\frac{\partial g}{\partial \eta})^T\} = \frac{2}{\sigma_n^2}\text{Re}[\mathbf{U}^H \mathbf{D} \mathbb{A}] \qquad (5.159)$$

$$E\{(\frac{\partial g}{\partial \mathbf{A}_{Im}})(\frac{\partial g}{\partial \eta})^T\} = \frac{2}{\sigma_n^4} E\{-\text{Im}[\mathbf{U}^H \mathbf{ww}^T \mathbf{D}^* \mathbb{A}^*] + \text{Im}[\mathbf{U}^H \mathbf{ww}^H \mathbf{D} \mathbb{A}]\}$$
$$\Rightarrow E\{(\frac{\partial g}{\partial \mathbf{A}_{Im}})(\frac{\partial g}{\partial \eta})^T\} = \frac{2}{\sigma_n^2}\text{Im}[\mathbf{U}^H \mathbf{D} \mathbb{A}] \qquad (5.160)$$

5.5. Lower Bounds of TDOA and DOA Estimation Error Variances

$$E\{(\frac{\partial g}{\partial \eta})(\frac{\partial g}{\partial \eta})^T\} = \frac{2}{\sigma_n^4}E\{\text{Re}[\mathbb{A}^H\mathbf{D}^H\mathbf{w}\mathbf{w}^T\mathbf{D}^*\mathbb{A}^*] + \text{Re}[\mathbb{A}^H\mathbf{D}^H\mathbf{w}\mathbf{w}^H\mathbf{D}\mathbb{A}]\}$$

$$\Rightarrow E\{(\frac{\partial g}{\partial \eta})(\frac{\partial g}{\partial \eta})^T\} = \frac{2}{\sigma_n^2}\text{Re}[\mathbb{A}^H\mathbf{D}^H\mathbf{D}\mathbb{A}] \quad (5.161)$$

The FIM of the deterministic parameters is obtained by $E\{(\frac{\partial g}{\partial \Omega})(\frac{\partial g}{\partial \Omega})^T\}$, where $\partial g/\partial \Omega = \partial g/\partial \{\sigma_n^2, \mathbf{A}_{Re}^T, \mathbf{A}_{Im}^T, \boldsymbol{\theta}, \boldsymbol{\tau}\}$. Finally, by using the results of [122], the CRB of the interesting parameters $\boldsymbol{\eta} = [\boldsymbol{\theta}, \boldsymbol{\tau}]^T$ (the DOAs and the relative time delays) for ULA-OFDM systems can be shown to be

$$\sigma^2_{\eta,CRB,ULA-OFDM} = \frac{\sigma_n^2}{2}\{\text{Re}[\mathbb{A}^H\mathbf{D}^H(\mathbf{I} - \mathbf{U}\mathbf{U}^\dagger)\mathbf{D}\mathbb{A}]\}^{-1} \quad (5.162)$$

It is worth mentioning that the output of (5.162) takes the following form

$$\sigma^2_{\eta,CRB,ULA-OFDM} = \begin{bmatrix} \{DOA\}_{L \times L} & \{DOA.TOA\}_{L \times L} \\ \{TOA.DOA\}_{L \times L} & \{TOA\}_{L \times L} \end{bmatrix} \quad (5.163)$$

where the diagonal elements of (5.162) are $[\sigma^2_{\theta_1}, \ldots, \sigma^2_{\theta_L}, \sigma^2_{\tau_1}, \ldots, \sigma^2_{\tau_L}]$, which represent our concern.

It is worth mentioning that the CRBs of time delay and DOA error variances are given in (5.162) for all effective paths. For wireless positioning, the CRBs of the first path ($\sigma^2_{\tau_1,CRB}, \sigma^2_{\theta_1,CRB}$) represent the most concern. The relation in (5.162) can be simplified to show the effect of the system parameters on $\sigma^2_{\tau_1,CRB}$ and $\sigma^2_{\theta_1,CRB}$ more obviously. To simplify the mathematical manipulations, it is recommended to assume a wireless channel of a single path. The following results are used: $\mathbf{d}^H_{\theta_1}\mathbf{d}_{\theta_1} = (2\pi f_c \rho \cos\theta_1/c)^2 NM(M^2 - 1)/12$, and $\mathbf{d}^H_{\tau_1}\mathbf{d}_{\tau_1} = (2\pi \Delta f)^2 MN(N^2 - 1)/12$, where $M > 2$. After some mathematical manipulations, it has been obtained

$$\sigma^2_{\theta_1,CRB} = \frac{6\sigma_n^2}{(2\pi f_c \rho \cos\theta_1/c)^2 NM(M^2 - 1)|\beta_1|^2}, \quad (5.164)$$

$$\sigma^2_{\tau_1,CRB} = \frac{6\sigma_n^2}{(2\pi \Delta f)^2 MN(N^2 - 1)|\beta_1|^2}. \quad (5.165)$$

Such as the previous, the following observations can be obtained from (5.164): the factor $(2\pi f_c \rho \cos\theta_1/c)$ represents the effect of the orientation of the ULA (array axis) with respect to the impinging wave direction. If the array axis is perpendicular to the impinging wave ($\theta = 0°$), the CRB of DOA estimation is the minimum value, but if the array axis is aligned to the same direction of the impinging wave ($\theta = \pm\pi/2$), the CRB tends to infinity. If the SNR $\gamma = |\beta_1|^2/\sigma_n^2$ increases, the error variance decreases. Increasing the number of antenna elements in the ULA M as well as the number of subcarriers N improves the accuracy of the DOA estimation. From (5.165), increasing the number of subcarriers including the subcarrier spacing as well as the number of antennas decreases the CRB error variance of the time delay estimation.

It is worth mentioning that to get the factorization in (5.130) for joint time delay and DOA estimation problem, it has been assumed that $\tau_1 \gg m\rho \sin\theta_1/c$ as described in Section 4.2. However, if the CRB is derived such as the previous for the DOA estimation only, the term Nf_c^2 in (5.164) should be replaced by $\sum_{k=0}^{N-1} f_k^2$, where $f_k = f_0 + k\Delta f$,

114 Chapter 5. Mobile Unit Position Estimation Based on TDOA and DOA Measurements

$k = 0, \ldots, N - 1$, and f_0 is the minimum frequency among subcarriers. However, the ratio between those two terms is one for 802.11ac bandwidths.

The CRB of joint time delay and DOA estimation depends on the ULA-OFDM system parameters. Those parameters include the SNR, the order of antenna array M, the system BW. To show the effect of the above parameters, assume the discussed parameter is variable and the remaining parameters are constants with the following values: the number of antennas M is 4, the system BW is 20 MHz, and the SNR is 20 dB. The carrier frequency f_c is 5.25 GHz, and the spacing of the antenna elements is the half wave length. Let us assume the number of paths is 3 with time delays $\tau = [20, 60, 90]$ ns and relative DOAs $\theta = [-5, 0, 20]$ degrees. The path gains are $\mathbf{A}_0 = [0, -9, -13]$ dB. To show the effect of SNR, Fig. 5.7 shows the STD of the first path time delay and the relative DOA with respect to SNR. To show the effect of a system BW, Fig. 5.8 shows the STD of the first path time delay and DOA with respect to the system BW. To show the effect of a number of antennas in the ULA, Fig. 5.9 shows the STD of the first path time delay and DOA with respect to the number of antennas.

As it is known, the multipath channel degrades the performance of the positioning system. This can be observed clearly from the CRB figures plotted in Sections 5.5.1 and 5.5.2, where a single path is used, and the CRB figures of this Section, where a multipath channel is used.

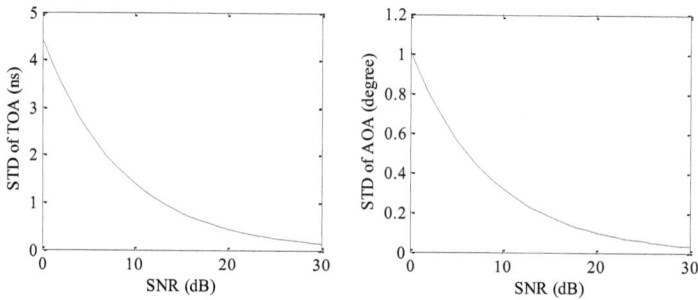

Figure 5.7: STD of the first path time delay and relative DOA estimates versus the SNR.

5.5. Lower Bounds of TDOA and DOA Estimation Error Variances

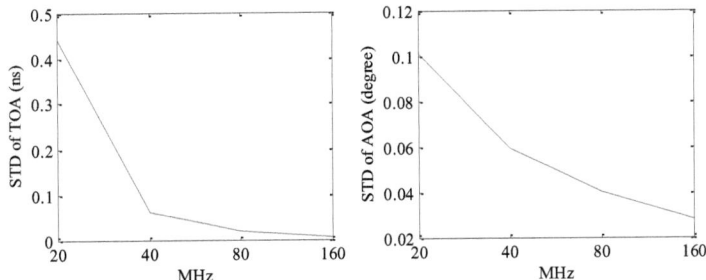

Figure 5.8: STD of the first path time delay and relative DOA estimates versus system BW.

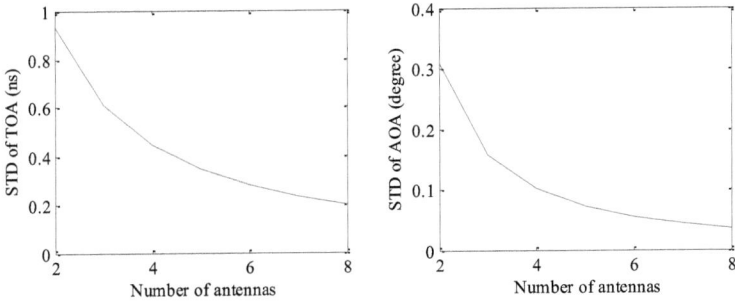

Figure 5.9: STD of the first path time delay and relative DOA estimates versus a number of antennas.

CHAPTER 6
Performance Evaluation Based on Channel Measurements

In this chapter, we present the capability of recent subspace-based algorithms to estimate the propagation time delays, and the relative DOAs associated with signals in a multipath communication channel for wireless indoor positioning using IEEE 802.11 standards. The 1-D and 2-D MP algorithms are applied in a new way to estimate the required parameters from the estimated S-CFR. The effect of temporal, spectral, and spatial diversity principles on the system performance will be presented. The performance of the proposed estimators of MU coordinates estimation will be investigated. To measure the performance of the proposed algorithms, the root mean square error (RMSE) of the estimated position will be calculated.

In the beginning, the necessary measurement tools will be presented. Then, the expected performance of using RSS methods will be confirmed. After that, the capability of 1-D MP algorithms to estimate the TDOA associated with signals in a multipath communication channel will be presented. The performance of the forward MP, and UMP, the forward-backward MP-Ex, and UMP-Ex, SBMP, and MBMP algorithms will be investigated and compared for TDOA estimation using OFDM systems. The accuracy and stability of various MP algorithms are investigated using 802.11a, 802.11n, and 802.11ac system parameters. The performance of multiple snapshot principle will be presented based on using multiple OFDM training symbols, a number of antennas, and frequency diversity. The effect of SNR on the performance will also be presented. The performance of using wider channel bandwidths is emphasized. To do that a number of experiments have been done. As an initial investigation for 1-D MP algorithms, a 1-D TDOA system has been built by using a direct connection between the transmitter and the receiver through a number of cables with different lengths, then using a LOS wireless channel with and without a reference signal, and then using a NLOS wireless channel. After that a 2-D wireless indoor positioning system has been built using four BSs. The performance of MU position estimation based on TDOA measurements in the XY plane will be presented.

The capability of 2-D MP algorithms will be presented to estimate the propagation time delays and the relative DOAs simultaneously using multi-antenna multi-carrier systems for wireless positioning. The performance of using 2-D MP, 2-D MP-Ex, 2-D UMP, and 2-D BMP will be investigated and compared. The performance of using multiple antennas and wideband orthogonal multi-carrier signals of 802.11ac will be presented. As an initial investigation for 2-D MP algorithms, LOS and NLOS experiments will be made. After that a 2-D wireless indoor positioning has been built by using four BSs. The performance of MU position estimation based on the TDOA and DOA measurements in the XY plane will be presented. A comparison will be made between the performance of using 1-D MP algorithms for TDOA estimation with spatial diversity and 2-D MP algorithms for hybrid TDOA and DOA estimation.

6.1 Measurement Tools

The real-time multi-carrier system can be achieved by using transmitter and receiver equipments, which can send and receive the real-time multi-carrier signal simultaneously. The analysis bandwidth of that type of equipments, which are available, is limited to 25 MHz. The real-time multi-carrier system can also be achieved using the channel sweeping at different frequencies. By using this principle, another type of equipments can be used to deal with the wide bandwidths of 802.11ac. This section presents the measurement tools that are used during this work for experimental investigation.

6.1.1 The Vector Signal Generator (Agilent MXG N5182A)

In our experiments, the RF vector signal generator from Agilent [126] shown in Fig. 6.1(a) acts as a transmitter for 802.11a, 802.11n, or 802.11ac frame. This signal generator covers the frequency range of 100 kHz to 8.5 GHz. The minimum and maximum output powers of MXG are -110 dBm and 13 dBm, respectively. The maximum analysis bandwidth is 25 MHz. The required parameters of MXG can be controlled via LAN interface using Standard Commands for Programmable Instruments (SCPI). Those parameters are the carrier frequency, the sampling frequency, the output power, and the buffer size. The baseband IQ data is generated using Matlab and sent to the signal generator via LAN connection. The signal generator saves the IQ data, applies the IF modulation and RF up conversion and then transmits the signal with the predefined parameters. MXG can be configured to transmit the same signal, which has been saved in the buffer, continuously.

6.1.2 The Antennas

The Tri-Band Rubber Duck antenna [127] is used. The frequency range of that type of antennas is from 2.4 GHz to 2.5 GHz, 4.9 GHz to 5.3 GHz, and 5.7 GHz to 5.8 GHz. The omni-directional antennas provide broad coverage and 3 dBi gain.

6.1.3 The Signal Analyzer (Agilent EXA N9010A)

The EXA signal analyzer from Agilent [128] shown in Fig. 6.1(b) acts as a receiver. This signal analyzer covers the frequency range of 9 kHz to 7 GHz. The maximum input power is 30 dBm, and the maximum analysis bandwidth is 25 MHz. It has an absolute sensitivity of -79.4 dBm and a dynamic range of 93.1 dB. The required parameters of EXA can be controlled using Matlab via LAN interface using SCPI commands. Those parameters are the carrier frequency, the analysis bandwidth, the time of recording, the mechanical attenuation, and the buffer size. The receiver downconverts the RF signal and generates the IQ baseband data which is sent back to the external PC for processing. By the way, the sampling rate of EXA is 45 MS/s, therefore, an appropriate filter should be used to convert the sampling rate to 20 MS/s. To do that the default filter of Matlab environment that is based on the Farrow structure has been used [129], [130].

6.1. Measurement Tools

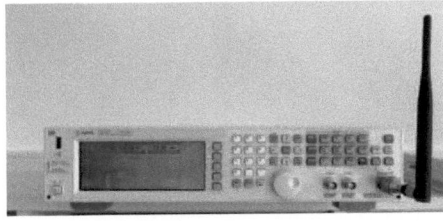

(a) Agilent MXG N5182A

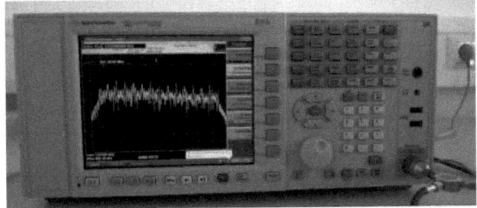

(b) Agilent EXA N9010A

Figure 6.1: The RF Vector Signal Generator (Agilent MXG N5182A) and the Signal Analyzer (Agilent EXA N9010A).

6.1.4 The Software Defined Radio Modules

The software defined radio (SDR) modules represented by the recent two versions from USRP (Universal Software Radio Peripheral) family are available in the laboratory, which can be used as a transceiver. Those are the USRP2 from Ettus [131] and NI USRP-2921 from National Instruments [132] as shown in Fig. 6.2. Both of them contain the RF front-end and the D/A and A/D converters. They are equipped with the XCVR2450 daughter-board, which has dual tunable bands of 2.4 GHz and 5 GHz. Although the XCVR2450 daughter-board has two ports, it does not support full-duplex mode or diversity mode. The maximum analysis bandwidth is 25 MHz. The maximum value of the overall gain of both analog and digital hardware receiver is 31.5 dB and of the transmitter is 25 dBm. The sample clock rate of A/D is 100 MS/s and of D/A is 400 MS/s. It is possible to synchronize a number of devices with an external reference clock and 1 Pulse Per Second (1PPS) signal. The high sample rate processing, like digital up and down conversion, takes place in the FPGA (Field Programmable Gate Array). The configurations and firmware are stored in a secure digital flash card in case of USRP2 while NI USRP is equipped with built-in memory. They can be connected to the host computer via Gigabit Ethernet card, which allows to send or receive 25 MS/s. The SDR environments of USRP2 and NI USRP have been integrated into the LabView and Matlab/Simulink tools. It is worth mentioning that the SDR packages of both tools can be used to receive at 20 MS/s, but they cannot be used to send at 20 MS/s in this time.

120   Chapter 6. Performance Evaluation Based on Channel Measurements

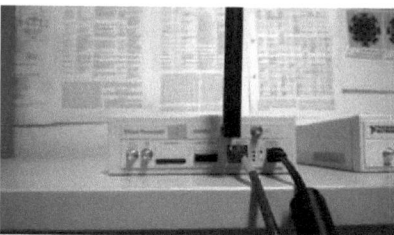

(a) USRP2

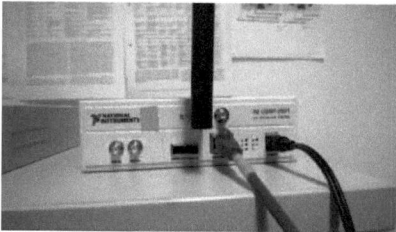

(b) NI USRP-2921

Figure 6.2: The SDR modules USRP2 from Ettus and NI USRP-2921 from National Instruments.

6.1.5 The Network Analyzer (Agilent ENA E5071C)

The network analyzer from Agilent [133] shown in Fig. 6.3 is used to measure the indoor CFR. ENA covers the frequency range of 9 kHz to 8.5 GHz. The maximum transmitted power of ENA is 10 dBm. The complex frequency channel response can be obtained by sweeping the channel at uniformly spaced frequencies. The real and imaginary parts of the forward transmission coefficient S21 can be measured and stored for further processing. A procedure of calibration should be followed before any experiment to reduce the effect of equipment characteristics on the measured data. The benefit of using ENA is that large bandwidths of 802.11 standards can be investigated. For example, in case of 160 MHz analysis bandwidth, each measurement recorded using ENA should cover a 160 MHz bandwidth, where the sampling interval should be set to the subcarrier spacing of 802.11 standards, $\Delta f = 312.5$ kHz. Therefore, the number of samples of the CFR is 512 samples, which is equal to the maximum IFFT/FFT order of 802.11ac.

6.1.6 The Wideband Time Domain Transmission (Agilent DCA 86100A)

The Agilent 86100A Infiniium [134] shown in Fig. 6.4 is a wide-bandwidth oscilloscope that also functions as a Digital Communication Analyzer (DCA) and a time-domain reflectometer/transmission (TDR/TDT). Since, the time delay estimation is the core of time-based wireless positioning, the wideband TDT is used to measure the propagation time delays of the system cables. It is also used to measure the time delay of the omni-directional antennas. To measure the time delay precisely, it is recommended to record the waveform

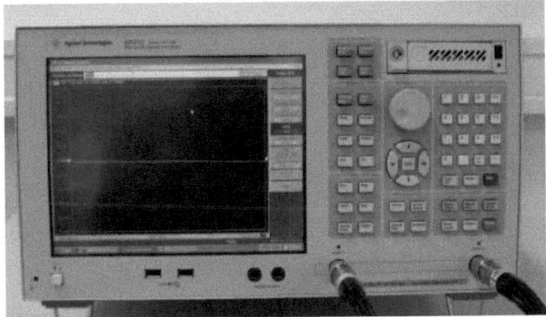

Figure 6.3: The Network Analyzer (Agilent ENA E5071C).

on DCA screen, and then calculate the time delay of a given setup, for example cables, from the derivative waveform to mitigate the problem of a large rise time, especially in case of using long cables.

6.2 RSS Techniques Investigation

As an initial investigation of radio signal characteristics, it is interested to discuss the expected performance from using the RSS for wireless positioning. In this section, the path loss model and the fingerprinting techniques presented in Section 2.6.1 are used to confirm why the RSS is not used in this work for high-resolution wireless positioning.

6.2.1 Investigation of Distance Estimation using Path Loss Model

To measure the capability of distance estimation using path loss model presented in Section 2.6.1.1, some experiments were performed. The RF Vector Signal Generator, Agilent MXG N5182A was used as the 802.11a transmitter. The SDR module represented by USRP2 was used as the 802.11a receiver. An overview of the interested part of 802.11a frame has been presented in Section 3.3.1. Omni-directional antennas were used in a LOS. The transmitter power of MXG was configured to 13 dBm, and the receiver gain of USRP2 was configured to 20 dB. The carrier frequency was 5.25 GHz.

Three experiments have been made, where in all of them, the RSS was measured using USRP2 at 10 positions. Those 10 positions have been distributed at distances from 1 m to 10 m with respect to the transmitter, MXG. The reference power was recorded at 1 m. The positions of USRP2 in the first and second experiments were in the middle of the lab as shown in Fig. 6.5, where the recording was in two different times (two days). For experiment three, the locations of USRP2 were in the left side of the lab as shown in Fig. 6.5. The estimated values of the mean path loss exponent n, and the measured powers at the reference distance, $d_0 = 1m$, have been summarized in Table 6.1. Fig. 6.6 shows the estimated distances with respect to the actual distances. The accuracy at distances larger than 5 m is very bad. The RSS is instantaneous inside the building and varies over the

122   Chapter 6. Performance Evaluation Based on Channel Measurements

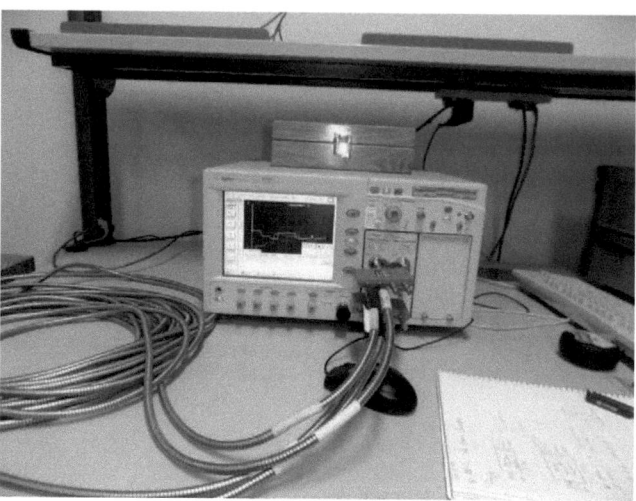

Figure 6.4: The Wideband Time Domain Transmission (Agilent DCA 86100A).

time, even at a fixed position. In general, using the path loss model for distance estimation based on the RSS cannot achieve high accuracy due to

1. The accuracy of using RSS decreases with greater distances due to the fact that the free space attenuation increases with the logarithm of the distance, where at somehow distance from the transmitter, the mean path loss exponent n has different value.

2. The value of RSS is changeable due to the radio channel impediments such as shadowing, and multipath effects as well as the orientation of the wireless device.

6.2.2 Investigation of Wireless Positioning using Fingerprinting

To get the principle and the required effort of the positioning based on the fingerprinting, a simple experiment has been made by using four receivers and one transmitter. The system parameters were the same of system parameters in Section 6.2.1, where four USRP2 modules were used as fixed BSs, and the signal generator (MXG) was used as a MU,

Table 6.1: The measured parameters of the path loss model.

Experiment number	$P(d_0)$ dBm	n
Exp.1 (time 1)	28.31	1.89
Exp.2 (time 2)	24.42	1.9
Exp.3	25.21	1.96

6.2. RSS Techniques Investigation

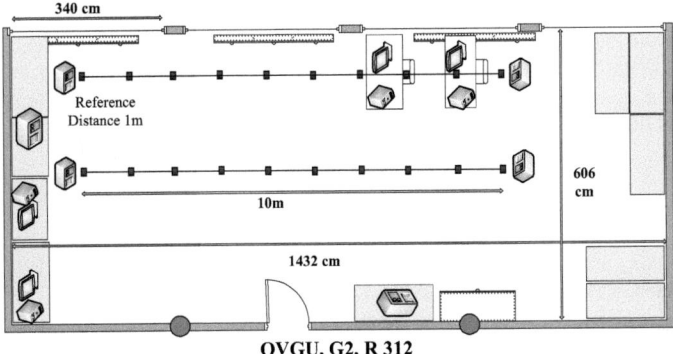

Figure 6.5: The environment of RSS technique investigation based on the path loss model.

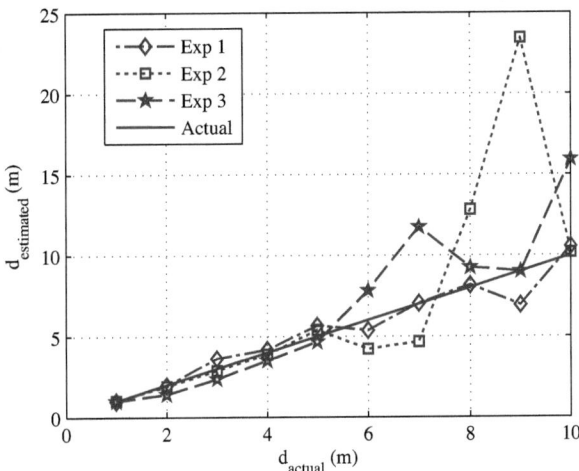

Figure 6.6: The estimated distances using the path loss model versus the actual distances.

which should be localized. The training sequence of IEEE 802.11a was used to measure the RSS in both offline and online modes. The radio map has been created by measuring the RSS of the transmitter at all receivers, where its location has been changed between points in the interested area. Fig. 6.7 shows the positions of fixed BSs (USRP2 modules), the points of radio map, and the position of MU, which should be estimated. At each point, the RSS of the transmitter has been measured 40 times, in another way, 40 OFDM frames have been recorded. The recorded radio map is presented in Table 6.2. The deviation of recorded RSS at MU position z can be obtained by using the Euclidean distance as

$$\sigma_{RSS}(dBm) = \sqrt{\sum_{i=1}^{4}(RSS_{ik} - RSS_{iz})^2} \qquad (6.1)$$

where RSS_{ik} is the measured RSS in dBm using USRP2 number i, $\{i = 1, \ldots, 4\}$ while the position of MU is at point a_k, $\{k = 1, \ldots, 9\}$, in the radio map. RSS_{iz} is the measured RSS using USRP2 number i while the position of MU is at point z, which should be estimated as shown in Fig. 6.7. The results of (6.1) using all points of radio map a_k is as follows

$$\sigma_{RSS}(dBm) = \begin{bmatrix} a_1 & a_2 & a_3 & a_4 & \mathbf{a_5} & a_6 & a_7 & a_8 & a_9 \\ 13.75 & 16.55 & 10.45 & 18.25 & \mathbf{3.89} & 12.36 & 19.24 & 6.49 & 18.78 \end{bmatrix}. \qquad (6.2)$$

From (6.2), the minimum deviation has been obtained at point a_5, therefore, the estimated coordinates of MU is the coordinates of point a_5.

It can be easily found that the radio map creation is a challenge. To locate the reference points inside the interested area, it needs a huge effort. The expected accuracy depends on the resolution of radio map creation process; the minimum distance between radio map points as shown clearly in Fig. 6.7. If the area of interest is large, the necessary number of BSs increases, and then the required effort of radio map creation and the size of database increase dramatically. Therefore, searching for the best match inside this large database requires a huge computational burden, and for any change in the environment, the creation of radio map should be repeated.

6.3 Simulation Results

6.3.1 Effect of Pencil Value

In this section, simulation results are presented to show the effect of the pencil parameter value. It was assumed that there are two paths with power 0.75 and 0.25 Watt. The time delay difference between them is 50 ns. The number of trials is 5000, and the SNR is 25 dB. The channel type is AWGN channel. Let us select the BW of 802.11ac or 802.11n to be 20 MHz. The first LTF is used to estimate the CFR; the number of CFR samples is $N = 57$ after doing an interpolation to mitigate the discontinuity at the carrier frequency. Fig. 6.8 shows the performance of all 1-D MP algorithms in terms of $-10log_{10}(RMSE)$ versus the pencil value, where the RMSE is in a nanosecond. It is clear that the best range of pencil value to be selected is between 16 and 40. To be more general, the pencil value should be selected around $N/3$ or $2N/3$. However, to reduce the complexity of SVD for 1-D MP algorithms, it should be selected around $2N/3$, as it is explained in Section 3.9.

6.3. Simulation Results

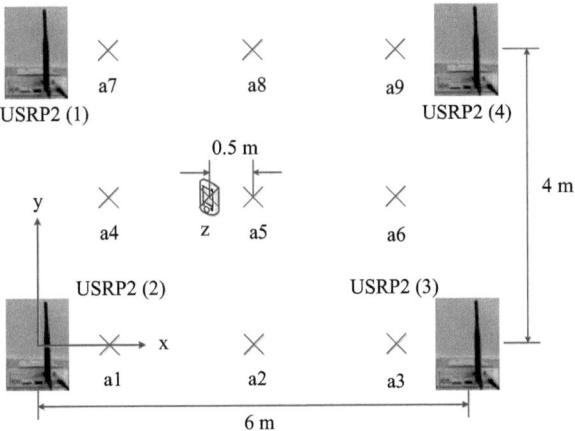

Figure 6.7: System setup of fingerprinting investigation

Table 6.2: The recorded RSS at radio map points with their coordinates.

Radio Map Points	(x,y) coordinates in meter	Measured RSS in dBm using			
		USRP2 (1)	USRP2 (2)	USRP2 (3)	USRP2 (4)
a1	(1,0)	6.50	23.50	11.39	3.42
a2	(3,0)	1.72	0.93	7.20	-2.20
a3	(5,0)	5.58	3.84	10.55	12.36
a4	(1,2)	17.57	14.7	-4.44	13.18
a5	(3,2)	5.44	9.55	5.42	8.33
a6	(5,2)	12.62	4.92	7.16	12.27
a7	(1,4)	20.36	6.37	2.76	15.08
a8	(3,4)	4.98	10.17	4.00	14.71
a9	(5,4)	-6.44	0.34	9.97	18.84
z	(2.5,2)	3.45	12.63	6.26	9.35

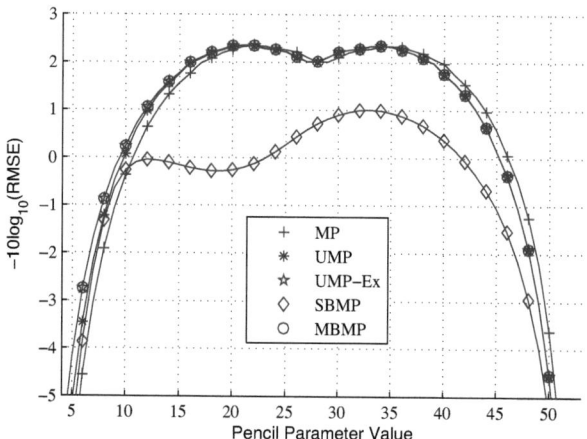

Figure 6.8: The performance of all 1-D MP algorithms at 20 MHz BW versus the pencil value.

6.3.2 Effect of SNR and Temporal and Frequency Diversities

In this section, simulation results are shown to illustrate the performance of using the 802.11ac BWs versus SNR, where the 1-D UMP algorithm is used, since it achieves the lowest complexity as it is shown in Fig. 3.7. It was assumed that there are two paths with gains 0 dB and -2.6 dB. The time-delay difference between them is 50 ns. The number of trials is set to 1000. At low SNR, the modified MDL criterion cannot detect the number of paths correctly, therefore, a threshold was used to configure large over estimation. Fig. 6.9 shows the $-10\log_{10}(RMSE)$ versus SNR for single snapshot with 20, 40, 80, 80+80, and 160 MHz BWs, and for 8 temporal snapshots with 20 MHz BW. It is clear that using wide BWs and multiple snapshots increases the accuracy at low SNR considerably. Fig. 6.10 shows the RMSE and number of real multiplications using 1-D UMP algorithm versus number of useful LTFs per OFDM frame, namely 1, 2, 4, 6, or 8, where 20 MHz BW and SNR of 20 dB were used. Results have been introduced in [54].

6.4 Measurement Systems Using 1-D MP Algorithms

In this section, the performance of various 1-D MP algorithms to estimate the propagation time delays is presented through some experiments. Those experiments include a propagation time delay difference between cables, a wireless channel with a reference signal, or LOS and NLOS wireless channels with spatial diversity by using a number of antennas.

6.4.1 Measurements Analysis using 20 MHz Bandwidth

The experimental results are presented to show the capability of MP, MP-Ex, UMP, UMP-Ex, SBMP, and MBMP algorithms to estimate TDOA using OFDM systems. The wide-

6.4. Measurement Systems Using 1-D MP Algorithms

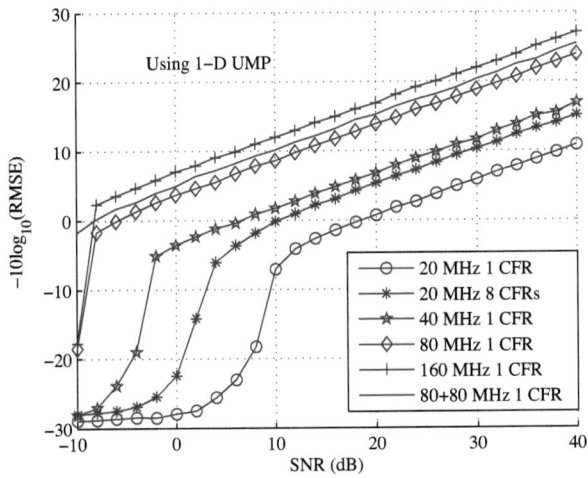

Figure 6.9: Comparison of estimation accuracy of various 802.11ac BWs using 1-D UMP algorithm versus SNR.

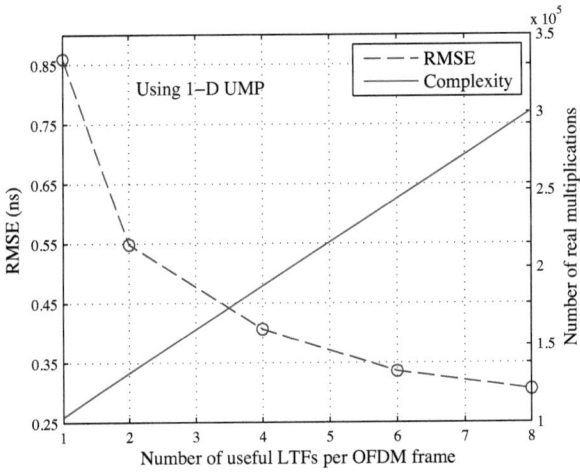

Figure 6.10: The RMSE and complexity versus number of snapshots with SNR of 20 dB, and BW of 20 MHz.

band TDT, Agilent 86100A, presented in Fig. 6.4 was used to measure the propagation time delays in some available cables. Two cables of lengths 405 cm and 2024 cm were used. The propagation time delay difference between cables is $dt_{21} = 65.82ns$ as shown in Fig. 6.11.

The RF Vector Signal Generator, Agilent MXG N5182A, was used to generate the 802.11a and 802.11n standards with output power of -20 dBm. The Signal Analyzer, Agilent EXA N9010A, was used as a receiver with a sampling frequency of $20MS/s$ and gain of 0 dB. A carrier frequency of 2.4 GHz was used. Due to the limitations of this equipment, the BW of 20 MHz only was used. For 802.11ac wide BWs, namely 40, 80, and 160 MHz, a network analyzer will be used as shown in the following sections. The transmitter and receiver were connected through the above mentioned cables in a power divider/power combiner configuration to build a 1-D TDOA estimation problem. The type of power splitters used in this experiment is ARRA A4200-4. Their frequency range is from 2 to 4 GHz, and the phase unbalance is 3 degrees. Fig. 6.12 shows the system configuration.

For further processing, 2000 OFDM frames are detected and recorded to estimate the corresponding CFRs using one LTS and one HT-LTF with Tx power of -20 dBm for a single snapshot principle. From Section 3.3, the frequency separation between pilots must be equal. Hence, an interpolation between the two subcarriers around the dc is used to mitigate the discontinuity of the CFR; consequently, N is equal to 53 and 57 in case of 802.11a and 802.11n, respectively. The pencil parameter was selected to be $P = 2N/3$, hence, $P = 35$ for 802.11a, and $P = 38$ for 802.11n as shown in Table 3.5.

To compare the performance of these algorithms, the RMSE and STD of the estimators are plotted in Figs. 6.13 and 6.14 for 802.11a and 802.11n, respectively. From Figs. 6.13 and 6.14, it can be noted that all MP algorithms resemble reliable accuracy and STD in the range of a few hundreds of picoseconds using one OFDM training symbol and 20 MHz BW. However, UMP-Ex and MBMP (when all invariances are used) achieve the same RMSE and STD, and the best performance. It is worth noting that converting to real computations removes part of the noise. The accuracy of all MP algorithms using 802.11n parameters is better than using 802.11a parameters. This is expected as 802.11n has larger useful BW; the number of occupied subcarriers in LTS of 802.11a and in LTF of 802.11n is 52 and 56, respectively, as it is presented in Table 3.1. Clearly, TDOA estimation accuracy strongly depends on the system BW. From Fig. 3.7, the complexity ratio of UMP-Ex to MBMP is around 0.70, and for UMP to MBMP is around 0.55. Therefore, regarding the accuracy and complexity, UMP-Ex is the best choice for TDOA estimation applications.

Fig. 6.15 shows a comparison between using 1 CFR and 2 CFRs with a more noisy signal, Tx power of -30 dBm. It is clear that using multiple snapshots increases the accuracy and decreases the variance of the estimators. It can also be noted that using multiple snapshots in UMP achieves some accuracy better than that in UMP-Ex, where the variance of UMP-Ex is smaller than that of UMP. However, using multiple snapshots increases the computational complexity. Results have been introduced in [53], [54].

6.4. Measurement Systems Using 1-D MP Algorithms

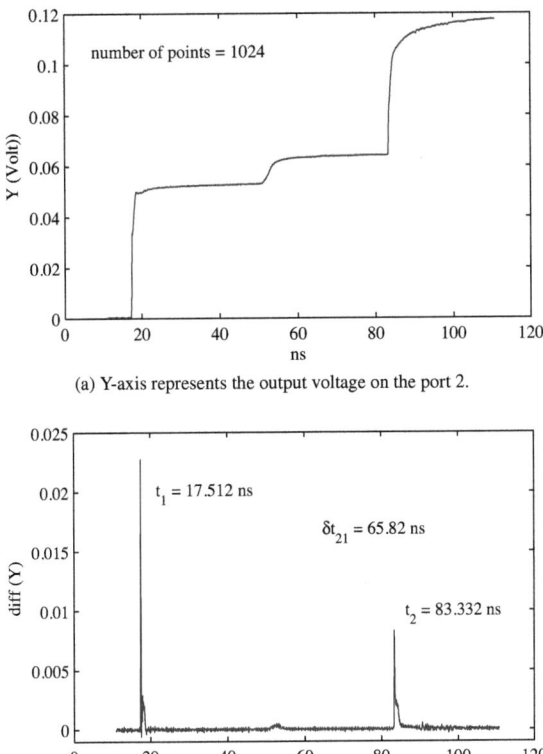

(a) Y-axis represents the output voltage on the port 2.

(b) Y-axis represents the derivative of the output voltage on the port 2.

Figure 6.11: The measured propagation time delays of path 1 and path 2 using the TDT.

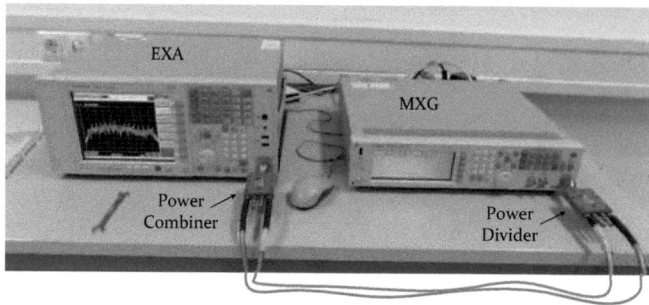

Figure 6.12: The measurement system using Vector Signal Generator and Signal Analyzer.

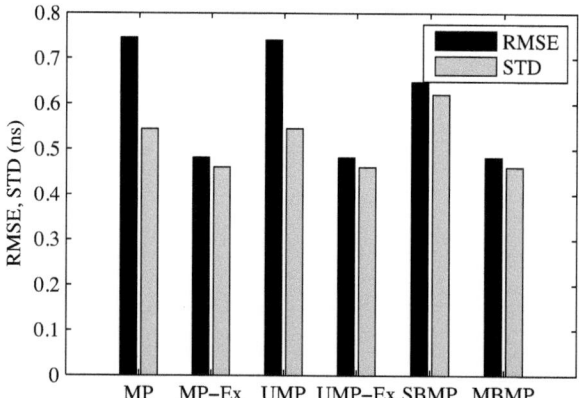

Figure 6.13: Comparison of estimation accuracy and stability of various MP algorithms using single snapshot with Tx power of -20 dBm and IEEE 802.11a parameters.

6.4. Measurement Systems Using 1-D MP Algorithms

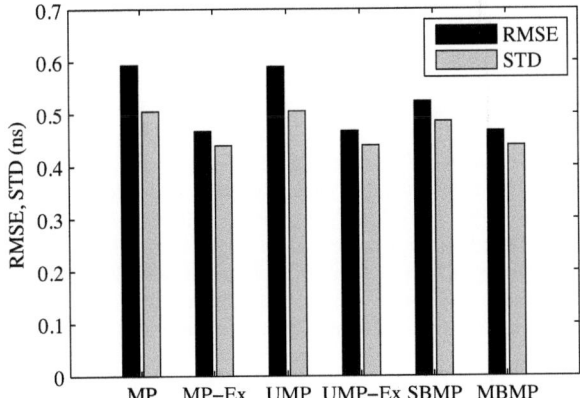

Figure 6.14: Comparison of estimation accuracy and stability of various MP algorithms using single snapshot with Tx power of -20 dBm and IEEE 802.11n parameters.

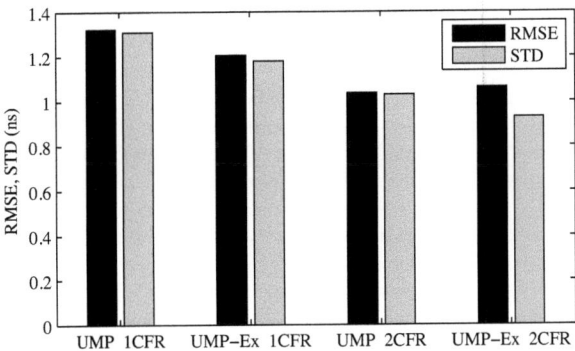

Figure 6.15: Comparison of estimation accuracy and stability of UMP and UMP-Ex algorithms using 1 and 2 snapshots with Tx power of -30 dBm.

6.4.2 TDOA Estimation using a LOS Wireless Channel and a Reference Signal

The second part of our experimental verifications is to investigate the TDOA estimation using the wider bandwidths of 802.11n and 802.11ac. The network analyzer, Agilent ENA E5071C, presented in Fig. 6.3 was used. It was used to measure the indoor frequency channel response in the frequency range of 802.11ac. A frequency range of 5170 MHz to 5330 MHz is used, which corresponds to a 160 MHz BW. The measurement system is shown in Fig. 6.16. To maintain the principle of TDOA estimation, the transmitter port and the receiver port of ENA were connected through cable 1 of length 405 cm, and cable 2 of length 701 cm plus a wireless channel in a power divider/power combiner configuration as shown in Fig. 6.16. The type of power splitters used in this experiment is ARRA A5200-2. Their frequency range is from 4 GHz to 8 GHz, and the phase unbalance is 3 degrees. To let the received power of direct path and shortest wireless path are comparable, an attenuator of 20 dB was used in the direct path as shown in Fig. 6.16. The TDT presented in Fig. 6.4 was used to measure the propagation time delays of those two cables with their necessary connections, where the time delay of path 1 (direct path) is 17.16 ns and of path 2 (wireless path) is 28.39 ns as shown in Fig. 6.17. Omni-directional antennas were used, which have 3 dBi gain and 0.668 ns time delay, measured using the TDT. The transmitted power of ENA was -5 dBm. Our goal is to estimate the time delay difference between direct estimated path, represented by cable 1, and first estimated wireless path as a 1-D TDOA problem.

The discrete samples of the CFR can be obtained using a multi-carrier modulation technique such as OFDM or channel sweeping at uniformly spaced frequencies. The frequency responses were collected at 8 LOS positions. The separation distances between the transmitter and receiver were arranged from 0 m to 7 m. The antenna height in both Tx and Rx was 152 cm. The real and imaginary parts of the forward transmission coefficient S21 were measured and stored for further processing. For each position, eight measurements were recorded for averaging purposes. Each measurement covers a 160 MHz bandwidth with a sampling interval set to 312.5 kHz. The number of CFR samples is 512, which is equal to the maximum IFFT/FFT order of 802.11ac.

Fig. 6.18 shows the RMSE and STD of each BW of 802.11ac using the single snapshot 1-D UMP-Ex algorithm. The parameters of each BW presented in Table 3.5 were used. It is clear that accuracy and STD in the range of a few hundreds of picoseconds can be achieved, and the accuracy and stability increase if the BW increases. However, from Table 3.4 and Fig. 3.7, if the BW increases, the computational complexity increases. Hence, to reduce the complexity of 160 MHz BW, it can be treated as two snapshots of 80 MHz BW, 80+80 MHz, as shown in Fig. 6.18, where a very small variance has been achieved. Results of this experiment have been published in [54].

6.4.3 Time Delay Estimation using a LOS Wireless Channel with Diversity Techniques

The goal of the experimental results presented in this section is to show the capability of 1-D matrix pencil algorithms including 1-D MP, 1-D MP-Ex, 1-D UMP, 1-D UMP-Ex, 1-D SBMP, and 1-D MBMP algorithms to estimate the propagation time delays using

6.4. Measurement Systems Using 1-D MP Algorithms

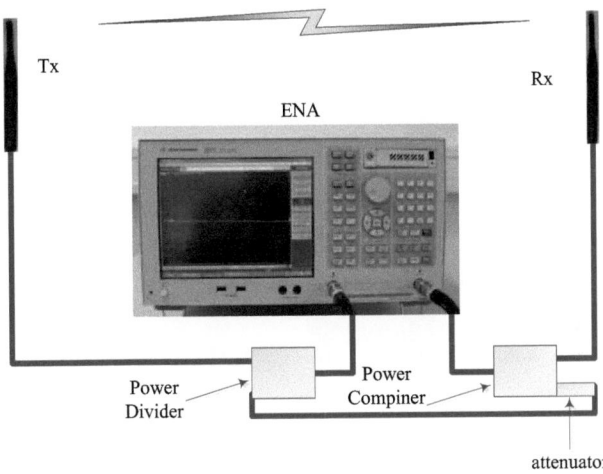

Figure 6.16: Frequency response measurement system using ENA with a reference signal.

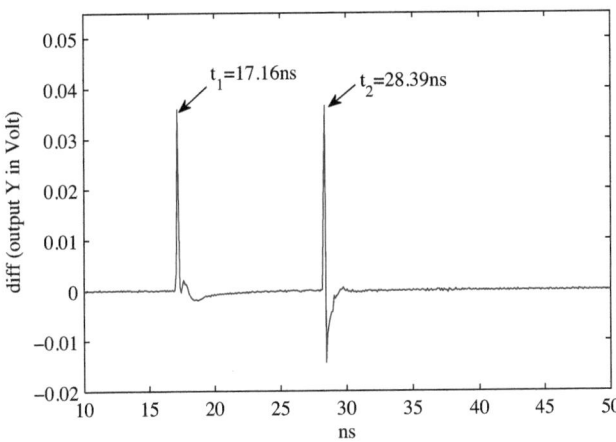

Figure 6.17: The measured propagation time delays of path 1 and path 2 using the TDT, where Y-axis represents the derivative of output voltage of port 2.

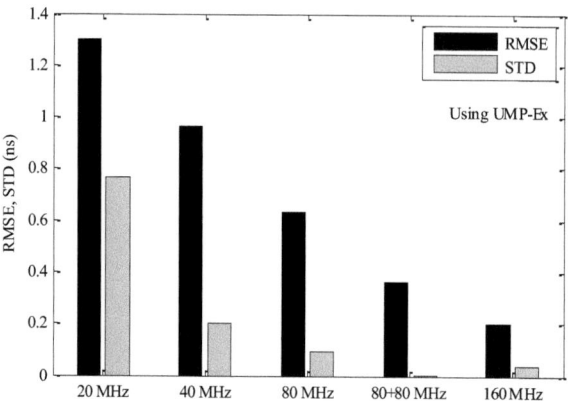

Figure 6.18: Comparison of estimation accuracy and stability of various 802.11ac BWs using single snapshot UMP-Ex algorithm.

multi-antenna multi-carrier systems. The effect of using the principle of multiple snapshots based on the spatial diversity and frequency diversity will be studied. The network analyzer, Agilent ENA E5071C, was used to measure the indoor CFR in the frequency range of 802.11ac. A frequency range of 5170 MHz to 5330 MHz is used. The system environment is shown in Fig. 6.19, and the inside view of the measurement locations in Fig. 6.20. The omni-directional antennas presented in the previous were used. The designed ULA presented in Fig. 4.3 was used at the receiver, where the maximum number of antenna elements is eight. As it has been mentioned in Section 4.2.1, the separation distance between antenna elements was designed to be 2.857 cm, calculated from the half wavelength of the carrier frequency 5.25 GHz. From Fig. 6.19, the transmitter antenna at the location point (150, 586.5) cm was connected to port 1 of ENA through cable 1 of length 12 m and time delay 45.64 ns, measured using the TDT. At the receiver side, one antenna from the antenna array elements was used to take the measurements at one time, which was connected to port 2 through cable 2 of length 1 m and time delay 4.79 ns, measured using the TDT. The remaining antenna elements in the antenna array were terminated by 50 Ohm loads. The connection between antenna elements and port 2 was changed manually. The coordinates of the antenna array center are (3, 11) cm. The transmitted power of the ENA is 10 dBm. Fig. 6.21 shows the measurement system using ENA and the designed antenna array.

The complex CFR of each antenna element can be obtained by sweeping the channel at uniformly spaced frequencies. The frequency responses were collected at LOS position. The antenna height in both Tx and Rx was 153 and 149.5 cm, respectively. The real and imaginary parts of the forward transmission coefficient S21 were measured and stored for further processing. For averaging purposes, 60 measurements were recorded during two days. Each measurement covers a 160 MHz BW with a sampling interval set to

6.4. Measurement Systems Using 1-D MP Algorithms

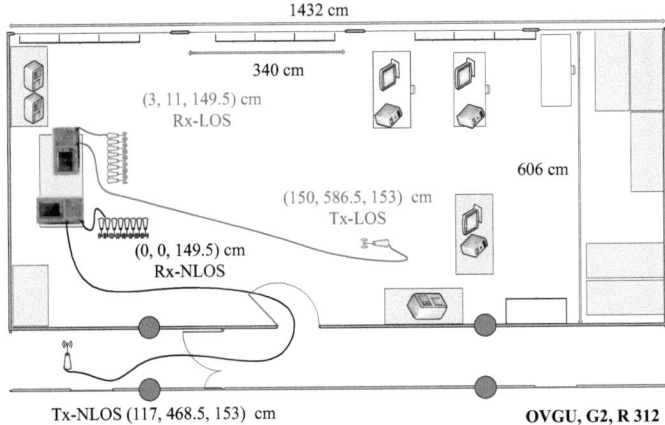

Figure 6.19: System environment in both scenarios of LOS and NLOS wireless channels between the transmitter and the receiver with ULA.

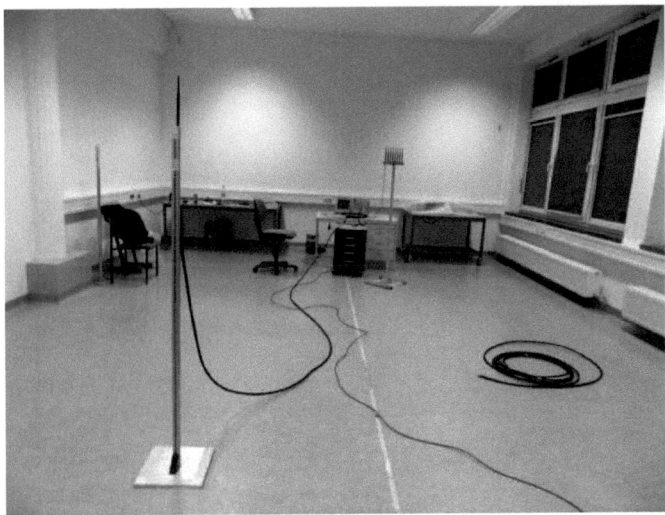

Figure 6.20: Inside view of measurement locations in case of LOS wireless channel.

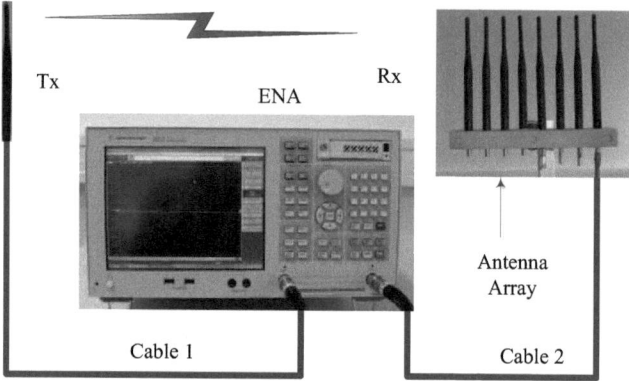

Figure 6.21: The measurement system using ENA and the designed antenna array.

312.5 kHz. The number of CFR samples is 512. In the following experiments using the network analyzer, a threshold was used to mitigate the large over estimation of a number of effective paths using the modified MDL criterion in (3.30) due to the low noise floor of the network analyzer. The number of effective paths will be limited to 12 paths if the result of the modified MDL is larger than that.

To show the performance comparison of all 1-D MP algorithms in case of LOS environment and the effect of using the spatial diversity, let us select the operating system BW to be 40 MHz and the number of antenna elements in the antenna array to be 3 antennas. The RMSE and STD of time delay estimation in nanoseconds are shown in Fig. 6.22. From Fig. 6.22, the accuracy and stability of time delay estimation of all 1-D MP algorithms are smaller than 700 ps (picoseconds). Also using the forward-backward principle in 1-D MP-Ex and 1-D UMP-Ex with the spatial diversity achieves the best performance, which is in the range of 220 ps. The 1-D UMP-Ex and 1-D MBMP achieve the same accuracy. Clearly, 1-D UMP-Ex is the best regarding the accuracy, stability, and complexity.

By using 1-D UMP-Ex algorithm, let us show the effect of increasing system bandwidth and using multiple snapshot principle based on the frequency diversity while M is 1. Fig. 6.23 shows the RMSE and STD in nanoseconds for all 802.11ac BWs and possible frequency diversities. In Fig. 6.23, 20+20, 40+40, and 80+80 MHz represent the frequency diversities of 40, 80, and 160 MHz BWs, respectively. The multiple snapshot principle was used as in (3.68). From Fig. 6.23, it is clear that using high BWs leads to high performance, for example at 160 MHz BW, the RMSE of time delay estimation is in the range of 30 ps. Using the frequency diversity in a multiple snapshot principle allows to use all measured CFR samples. Consequence, a huge reduction in the complexity can be achieved as shown in Fig. 3.7 with a little bit reduction in the performance. For example, in Fig. 6.23, by using 40+40 MHz BW rather than 80 MHz BW or 80+80 MHz BW rather than 160 MHz BW, a comparable performance can be achieved.

Finally, it is interesting to show in this experiment also the comparison between

6.4. Measurement Systems Using 1-D MP Algorithms

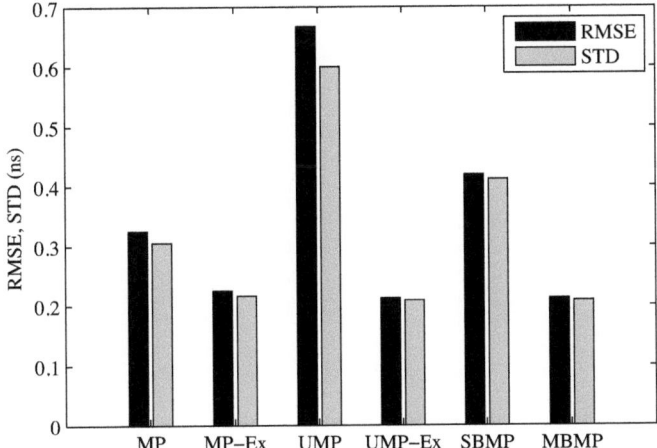

Figure 6.22: Comparison of time delay estimation accuracy and stability of various 1-D MP algorithms using 40 MHz BW and 3 antenna elements in a LOS environment.

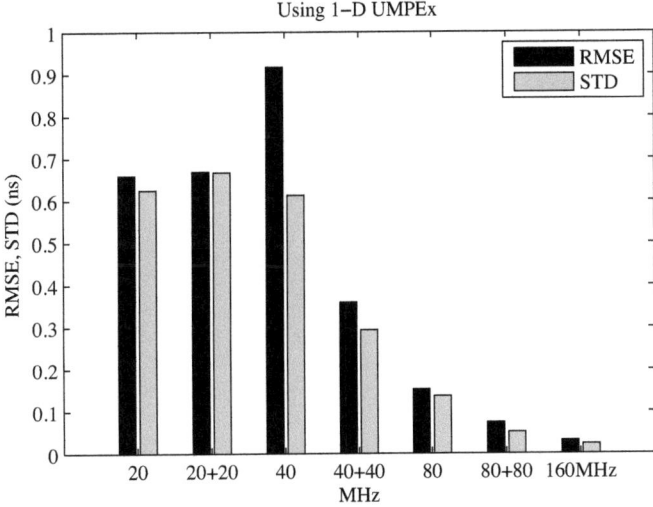

Figure 6.23: Comparison of estimation accuracy and stability of various 802.11ac BWs and possible frequency diversities using UMP-Ex algorithm and single antenna.

achieved performance from increasing system BW and spatial diversity order. Fig. 6.24 shows the performance of time delay estimation of 1-D UMP-Ex using various 802.11ac BWs and a number of antenna elements, which could be 1, or 2,..., or 8 plotted from left to right for each BW. The maximum number of antenna elements in 802.11ac is eight antennas. To make a reliable conclusion, let us repeat the previous LOS measurement at different positions for Tx and Rx inside the environment of Fig. 6.19. The maximum order of ULA at Rx is 6. Fig. 6.25 shows the performance of using various 802.11ac BWs and different orders of ULA. From Figs. 6.24 and 6.25, using spatial diversity for 1-D MP algorithms in LOS environments improves the accuracy of time delay estimation slightly, but the achieved performance is not like that of increasing system BW. In addition, using high orders of spatial diversity at high BWs is approximately useless in LOS channels. It is worth mentioning that the bars in Fig. 6.24 and Fig. 6.25 are not smooth due to the small number of measurements.

6.4.4 Time Delay Estimation using a NLOS Wireless Channel with Diversity Techniques

The system parameters used in the previous section of LOS environment are used here for a NLOS environment. The system environment is shown in Fig. 6.19, and the inside view of the measurement locations in Fig. 6.26 . As shown in Fig. 6.19, the coordinates of the transmitter antenna in the corridor are (117, 468.5, 153) cm, and the coordinates of the antenna array center in the lab are (0,0,149.5) cm. At the receiver side in the lab, the number of antenna elements in the ULA could be 4, 6, or 8, where one antenna from the antenna array elements was used to record the measurements at one time. The remaining antenna elements in the antenna array were terminated by 50 Ohm loads. The connection between antenna elements and port 2 was changed manually.

The frequency responses were collected at a NLOS position. There is a wall between Tx and Rx of thickness 13 cm. The real and imaginary parts of the forward transmission coefficient S21 were measured and stored for further processing. The number of recorded measurements is 60, collected during two days for averaging purposes. Such as the previous experiment, each measurement covers a 160 MHz BW with a sampling interval set to 312.5 kHz.

To show the performance comparison of all 1-D MP algorithms in case of a NLOS environment and the effect of using the spatial diversity, the operating system BW was selected to be 40 MHz and the array order to be 3. The RMSE and STD of time delay estimation are shown in Fig. 6.27. The accuracy and stability of all 1-D MP algorithms are smaller than 1.75 ns. If we compare between the performance presented in Fig. 6.27 of NLOS and that of LOS presented in Fig. 6.22, a stability degradation occurs in case of NLOS by amount of 1 ns. However, if the direct path is still detected in the received signal, the performance can be improved. For example, if the system BW has been increased to 160 MHz BW and the number of antenna elements has been increased to 8 elements, a huge improvement can be achieved as shown in Fig. 6.28. From Fig. 6.28, all 1-D MP algorithms achieve accuracy and stability smaller than 30 ps rather than 1-D UMP, where its RMSE and STD are 54 ps and 65 ps, respectively. The achieved RMSE and STD of the forward and backward 1-D UMP-Ex and MBMP are 15 ps and 22 ps, respectively.

Finally, Fig. 6.29 shows the comparison between the achieved performance to es-

6.4. Measurement Systems Using 1-D MP Algorithms

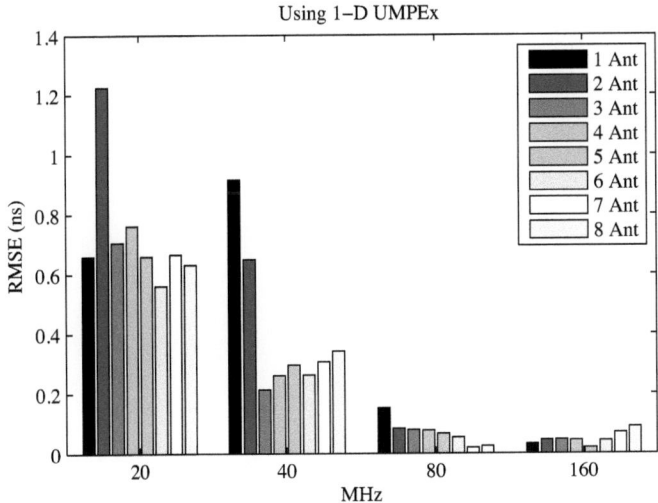

Figure 6.24: Comparison of time delay estimation accuracy of 1-D UMP-Ex using various 802.11ac BWs and different orders of ULA (1 to 8) plotted from left to right for each BW in a LOS environment.

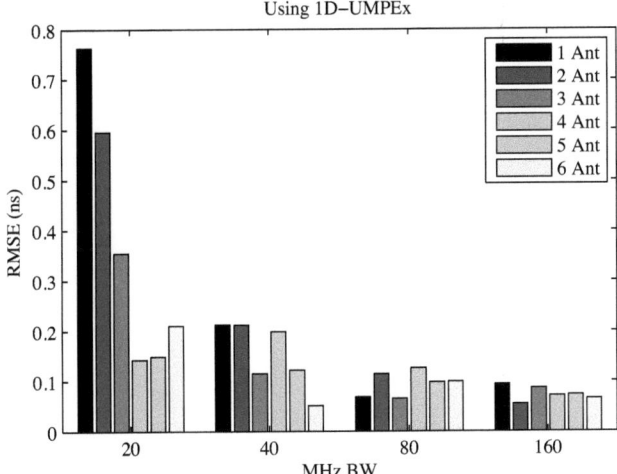

Figure 6.25: Comparison of time delay estimation accuracy of 1-D UMP-Ex using various 802.11ac BWs and different orders of ULA (1 to 6) plotted from left to right for each BW in a LOS environment.

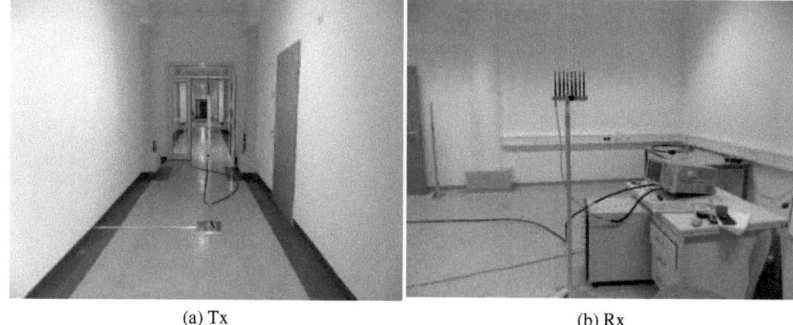

(a) Tx   (b) Rx

Figure 6.26: Inside view of measurement locations in case of NLOS wireless channel.

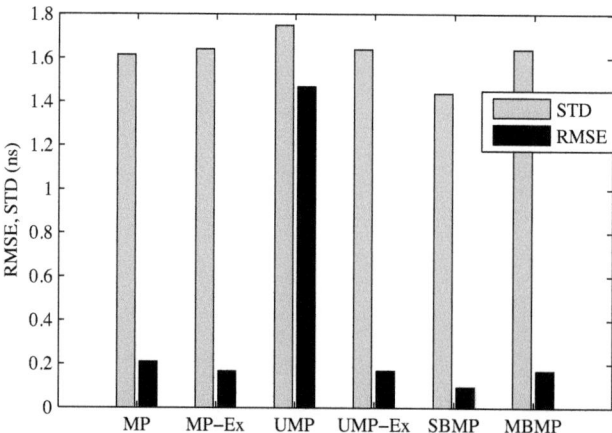

Figure 6.27: Comparison of time delay estimation accuracy and stability of various 1-D MP algorithms using 40 MHz BW and 3 antenna elements in a NLOS environment.

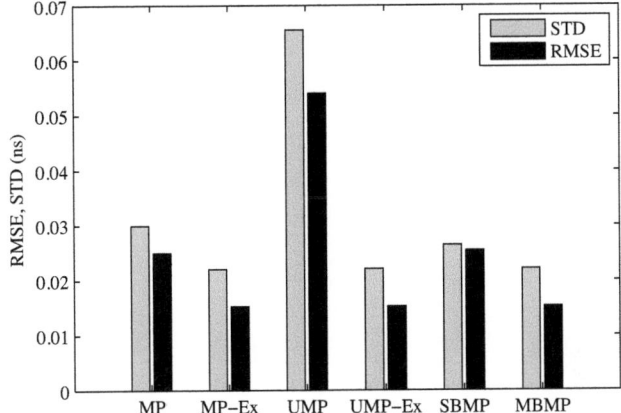

Figure 6.28: Comparison of time delay estimation accuracy and stability of various 1-D MP algorithms using 160 MHz BW and 8 antenna elements in a NLOS environment.

timate time delays using 1-D UMP-Ex by increasing system BW and spatial diversity order. From Fig. 6.29, it can be observed for NLOS scenarios that it is recommended to use wide BWs with spatial diversity to estimate the direct path precisely.

6.5 Measurement Systems Using 2-D MP Algorithms

In this section, some experimental results are presented to show the capability of 2-D MP algorithms including 2-D MP, 2-D MP-Ex, 2-D UMP, and 2-D BMP algorithms to estimate the propagation time delays and the relative DOAs of a wireless channel using multi-antenna multi-carrier systems. The previous two experiments presented in Fig. 6.19 of LOS and NLOS using 1-D MP algorithms with spatial diversity will be repeated here using 2-D MP algorithms.

6.5.1 Joint Time Delay and DOA Estimation using a LOS Wireless Channel

The system parameters and configurations presented in Section 6.4.3 and shown in Fig. 6.19 and Fig. 6.21 are also used here. At the receiver, the number of antenna elements in the ULA is eight. Fig. 6.30 shows the comparison of time delay and DOA estimation accuracy of 2-D MP algorithms, where the system BW is 40 MHz and the number of antenna elements is 3. From Fig. 6.30, the RMSE of time delay estimation is smaller than 200 ps and the RMSE of DOA estimation is smaller than 2.54 degrees. By using the 2-D UMP algorithm, Fig. 6.31 shows the RMSE of both time delay and DOA estimation in case of using 20, 40, 80, or 160 MHz BW and in case of using 3, 4, 6, or 8 antenna elements plotted from left to right for each BW. From Fig. 6.31 of a LOS environment,

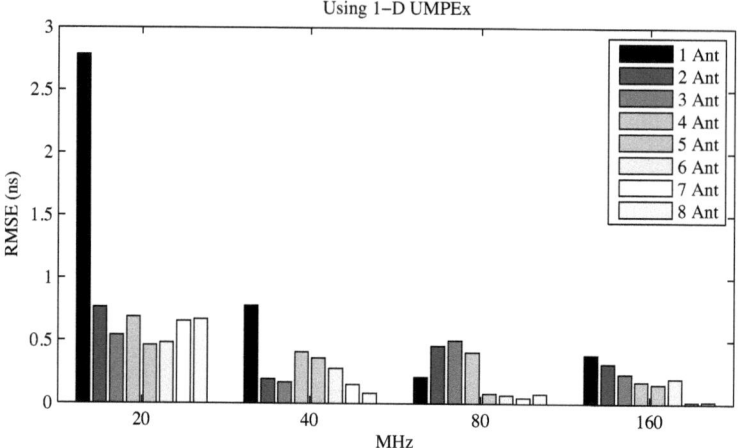

Figure 6.29: Comparison of time delay estimation accuracy of 1-D UMP-Ex using various 802.11ac BWs and antenna array order 1 to 8 plotted from left to right for each BW in a NLOS environment.

using a number of antennas improves the accuracy of DOA estimation. The accuracy of time delay estimation is very high in a LOS environment.

If we compare between the RMSE of time delay estimation of 1-D UMP-Ex presented in Figs. 6.24 and 6.25 and of 2-D UMP presented in Fig. 6.31, it can be noted that in general the accuracy of 1-D UMP-Ex is a little bit better than that of 2-D UMP in a LOS channel. However, in case of using 2-D UMP, an extra information is available, which is the DOA. Therefore, a precise comparison will be given later, where a full 2-D wireless indoor positioning will be built and all estimated values either time delays or DOAs are used for MU coordinates estimation.

6.5.2 Joint Time Delay and DOA Estimation using a NLOS Wireless Channel

The system parameters and configurations presented in Section 6.4.4 and shown in Fig. 6.19 and Fig. 6.21 are also used here, where the number of antenna elements in the antenna array could be 3, 4, 6, or 8. By using 40 MHz BW and 3 antenna elements, Fig. 6.32 shows the RMSE of both time delay and DOA estimations of various 2-D MP algorithms. The real MP algorithms, represented by 2-D UMP and 2-D BMP, achieve RMSE of time delay estimation in the range of 910 ps and RMSE of DOA estimation in the range of 9.1 degrees. To improve the performance in the NLOS environment, let us increase the system BW to 80 MHz and the number of antenna elements to 8. Fig. 6.33 shows the RMSE of time delay and DOA estimations. In case of real MP algorithms, the achieved RMSE of time delay estimation is in the range of 60 ps and of DOA estimation is in the range of 0.2 degree. Fig. 6.34 shows the performance of 2-D UMP using various

6.5. Measurement Systems Using 2-D MP Algorithms

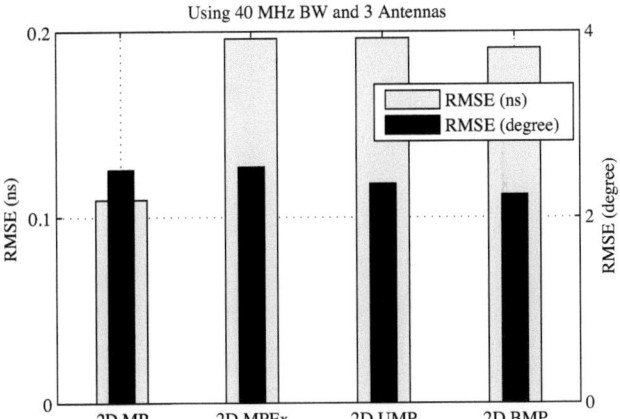

Figure 6.30: Comparison of time delay and DOA estimation accuracy of various 2-D MP algorithms using 40 MHz BW and 3 antenna elements in a LOS environment.

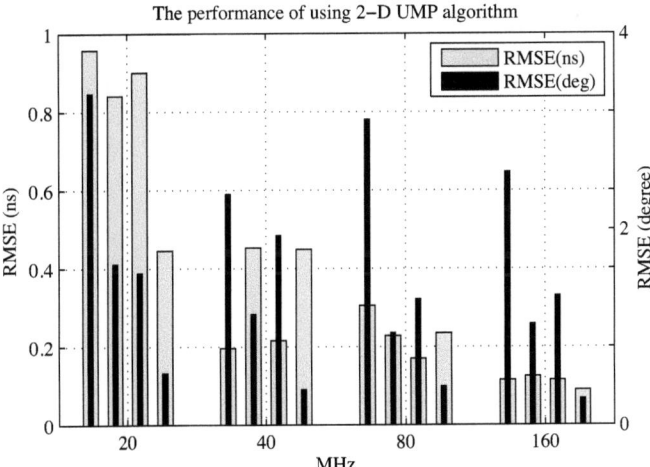

Figure 6.31: Comparison of time delay and DOA estimation accuracy of 2-D UMP using various 802.11ac BWs and a number of antenna elements (3, 4, 6, or 8) plotted from left to right for each BW in a LOS environment.

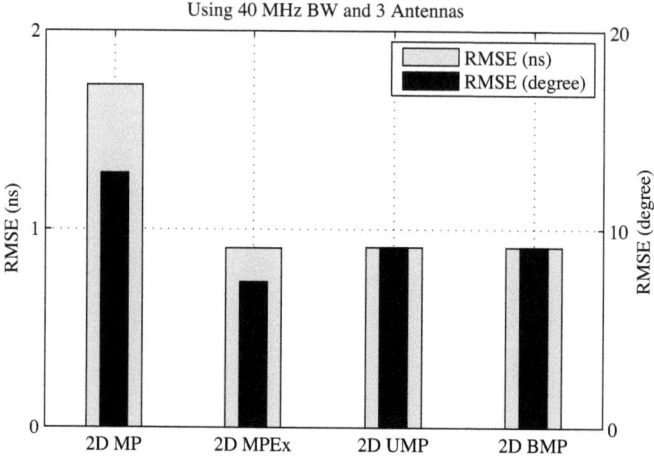

Figure 6.32: Comparison of time delay and DOA estimation accuracy of various 2-D MP algorithms using 40 MHz BW and 3 antenna elements in a NLOS environment.

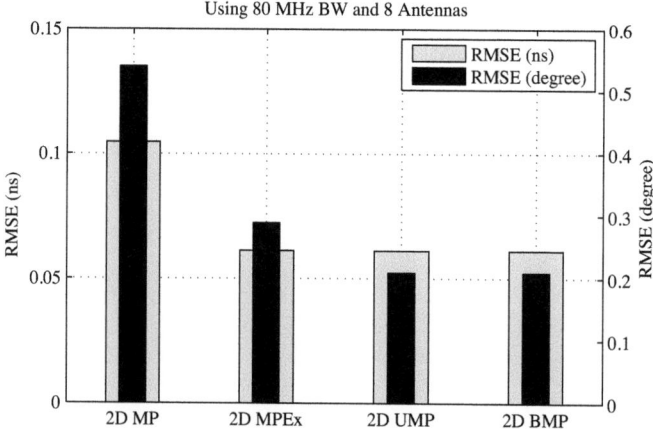

Figure 6.33: Comparison of time delay and DOA estimation accuracy of various 2-D MP algorithms using 80 MHz BW and 8 antenna elements in a NLOS environment.

802.11ac BWs, and a number of antenna elements which could be 3, 4, 6, or 8 plotted from left to right for each BW. If the BW and / or the number of antennas increase(s), the accuracy of time delay and DOA estimation increases. It should be noted that if the number of subcarriers increases (BW), the necessary number of antennas for accurate DOA decreases, which can reduce the receiver complexity.

From the previous, we can conclude that using multi-antenna multi-carrier principles can successfully enhance the dimensionality of the signal subspace for joint time delay and DOA estimation; it represents a robust technique versus multipath channel fading. The 2-D UMP is the best choice for wireless positioning regarding complexity and accuracy. Based on the NLOS experimental results, accuracy in the range of a few tens of picoseconds, and a fraction of one degree could be achieved. It is worth mentioning that using different wall may lead to different results, where each wall has different characteristics. Results have been presented in [57].

6.6 Measurement Analysis for 2-D Wireless Indoor Positioning

In this section, the problem of highly resolving the channel profile parameters of multipath signals will be investigated using a single transmitter and a number of receiving BSs for 2-D wireless indoor positioning based on the TDOA or hybrid TDOA and DOA observations. Fig. 6.35 shows the test environment used in this work. To estimate XY coordinates of the transmitter based on the received signal characteristics, four BSs are used, where two BSs are located in the Lab and in LOS with the transmitter while the other two BSs are located in the corridor and in NLOS with the transmitter as shown in Fig. 6.35. A number of antennas could be used at each BS for space diversity or DOA estimation.

The experimental results are presented to show the capability of using the enhanced 1-D and 2-D MP algorithms to estimate the propagation time delays, and the time delays with the relative DOAs, respectively, using multi-antenna multi-carrier systems. The network analyzer was used. It was used to measure the indoor CFR in the frequency range of 802.11ac. The measurement system is shown in Fig. 6.21. Omni-directional antennas were used, which have 3 dBi gain and 0.668 ns time delay, measured using the wideband TDT. At the receivers, the number of antenna elements in the ULA could be 1, 2,..., or 6 for each BS. The separation distance between antenna elements was designed to be the half wavelength of the carrier frequency 5.25 GHz.

The XY coordinates of receivers and transmitter (in cm) are as follows: BS1 (146.2, -172.6), BS2 (841.6, -213.2), BS3 (907.7, 338.2), BS4 (19.1, 333.4), and Tx (462.1, 161.4), respectively, where the coordinates of each BS are the coordinates of its antenna array center. Fig. 6.36 shows the locations of the four BSs and the MU in the XY plane with the relative angles. The height of antenna arrays in all BSs is 149.5 cm, and the height of Tx antenna is 153 cm. The transmitter antenna at the location point in the lab was connected to port 1 of ENA through cable 1 of length 100 cm and time delay 4.79 ns, measured using the TDT. At the receiver sides, one antenna from the antenna array elements was used to take the measurements at one time, which was connected to port 2 through cable 2 of length 1230 cm and time delay 46.807 ns, measured using the TDT. The

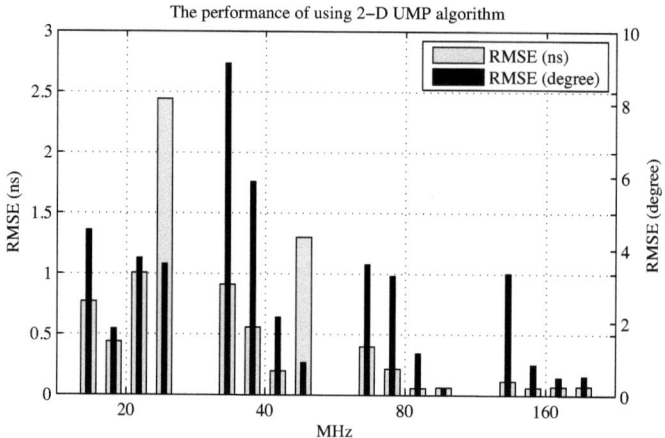

Figure 6.34: Comparison of time delay and DOA estimation accuracy of 2-D UMP using various 802.11ac BWs and a number of antenna elements (3, 4, 6, or 8) plotted from left to right for each BW in a NLOS environment.

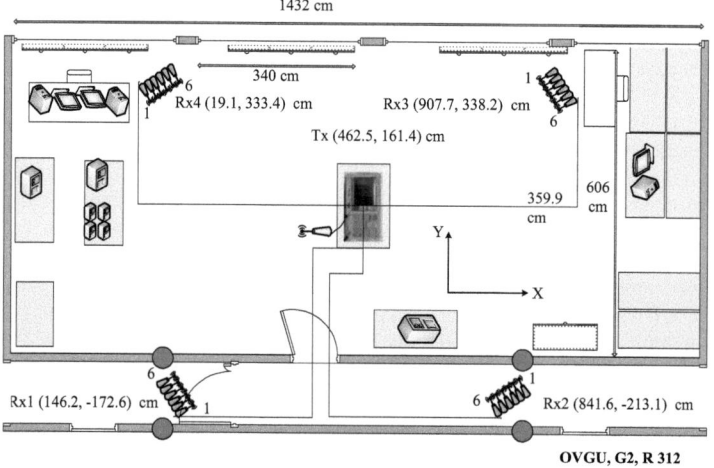

Figure 6.35: The lab and corridor used as test environment.

6.6. Measurement Analysis for 2-D Wireless Indoor Positioning

remaining antenna elements in the antenna array were terminated by 50 Ohm loads. The connection between array elements and port 2 was changed manually. The transmitted power of the ENA was 10 dBm.

The CFRs were collected at LOS and NLOS positions. There is a wall between the transmitter and the two BSs in the corridor of thickness 13 cm as shown in Fig. 6.35. For each Rx antenna, 310 measurements were recorded for averaging purposes during eight days and at different times during the day, the human body is not present. Each measurement covers 80 MHz BW with a sampling interval set to 312.5 KHz. The number of CFR samples is 256. Such as the previous, a threshold was used to mitigate the large over estimation of a number of effective paths using the modified MDL. The main reason of recording data at different times is that accuracy with the same deviation is obtained if the indoor channel is measured in a very small time interval, which doesn't represent the actual performance. As a result, the CFRs should be recorded at different times through the day. The relationship between the accuracy and the time of recording in different environments using many devices is still a future work.

In the following, we will present and compare between the achieved performance by using 1-D UMP-Ex and 2-D UMP algorithms for channel profile parameters estimation, and the presented algorithms in Table 5.1 for MU coordinates estimation. To estimate the transmitter position using TDOA observations, the following estimators will be used: ILS and W-ILS estimators of Section 5.1.1, DAC-LS estimator of Section 5.1.2, Chan estimator of Section 5.1.3, and the proposed LS and W-LS estimators of Section 5.4.1. To estimate the transmitter position using DOA observations, the following estimators will be used: LI estimator of Section 5.2.1, LS and W-LS estimators of Section 5.2.2, SLS and W-SLS estimators of Section 5.2.4, and DAC-LI and A-DAC-LI estimators of Section 5.2.3. To combine TDOA and DOA observations together, the following estimators will be used: Hybrid-DAC estimator of Section 5.3.1, Hybrid-W estimator of Section 5.3.2, and the proposed Hybrid-WLS estimator of Section 5.4. The impact of the number of antennas M for each array and the system bandwidth will be analyzed.

To show the performance of the above methods, the RMSE and the distribution of the position error represented by the cumulative distribution function (CDF) will be used. The following criterion will be used to evaluate the distribution of the position error

$$e_{\mathbf{z}} = \|\hat{\mathbf{z}} - \mathbf{z}\| = \sqrt{(\hat{x} - x)^2 + (\hat{y} - y)^2} \tag{6.3}$$

where $e_{\mathbf{z}}$ represents the distance between the estimated position and the actual position of the MU.

6.6.1 Using 1-D MP Algorithms for 2-D Wireless Indoor Positioning

6.6.1.1 Position RMSE versus System Bandwidths

In this section, the performance of using 1-D UMP-Ex algorithm and the introduced MU position estimators in Chapter 5 that are based on TDOA observations will be presented. Fig. 6.37 shows the RMSE of the estimated XY coordinates in cm, where a single antenna per each BS was used. It can be noted that at lowest system parameters such as 20 MHz BW and single antenna, using the variance of measurements for weighting to estimate MU coordinates improves the performance of XY coordinates estimation, especially if

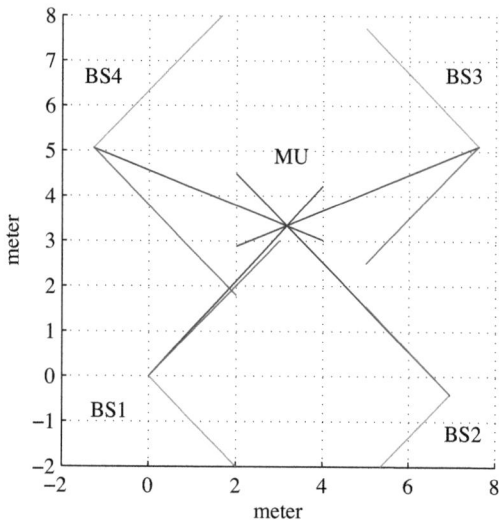

Figure 6.36: The locations of base stations and MU presented the relative angles.

the TDOA observations are noisy. In additional to the lowest complexity of the proposed W-LS estimator, which requires only a single iteration, it achieves a modest performance improvement over the other estimators. Chan estimator at those lowest system parameters is the worst estimator while the achieved RMSE is 553 cm at 20 MHz BW and M = 1, which is outside the scope of Fig. 6.37. Also it is obviously that using high BWs improves the performance of MU position estimation considerably. Fig. 6.38 and 6.39 shows the STD of the estimated X and Y coordinates in cm while the various 802.11ac BWs with a single antenna are used. It is clear that the STD of Y estimates is more than that of X estimates due to the non-symmetry in the locations of BSs as shown in Fig. 6.36. Therefore, the location of BSs inside the building should be designed properly. Results have been introduced in [55].

6.6.1.2 Position RMSE versus Number of Antennas

In this section, we need to study the effect of increasing number of antennas in all BSs for spatial diversity. The RMSE of the estimated XY coordinates by using 20, 40, and 80 MHz BWs, and a number of antennas $M = 1, \ldots, 6$ for each BS is shown in Fig. 6.40, in which the W-LS estimator is used. Clearly, a huge performance improvement has been obtained by using high BWs with spatial diversity. For example, the RMSE is just 1.64 cm in case of using 80 MHz BW and $M = 3$. Fig. 6.41 shows the distribution of the estimated XY coordinates in the XY plane while 40 MHz BW and 3 antennas in the ULA for each BS were used. The W-LS estimator was used to estimate the MU coordinates.

Let us now plot the CDF of error distance e_z between the estimated MU position and the actual position using (6.3). Fig. 6.42 shows the CDFs of the position error e_z

6.6. Measurement Analysis for 2-D Wireless Indoor Positioning

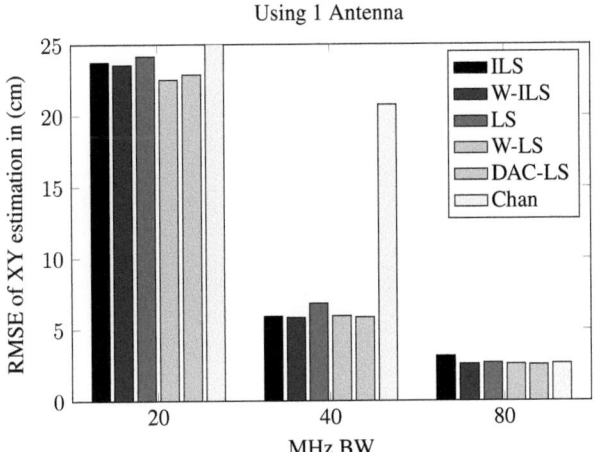

Figure 6.37: Comparison of XY coordinates estimation accuracy of various 802.11ac BWs and MU position estimators, where $M = 1$.

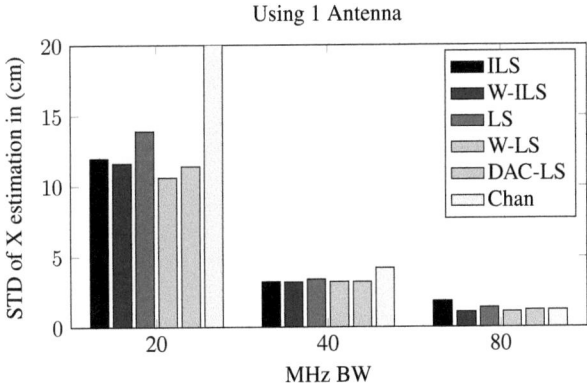

Figure 6.38: The STD of X coordinate estimates in cm.

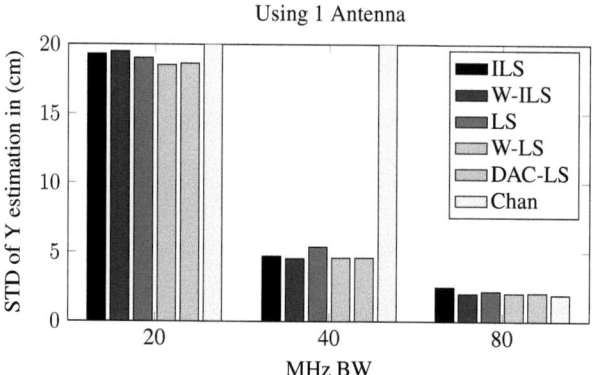

Figure 6.39: The STD of Y coordinate estimates in cm.

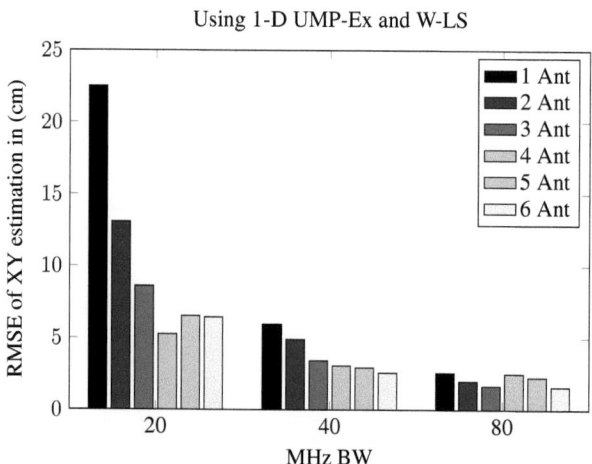

Figure 6.40: Comparison of XY coordinates estimation accuracy of various oder of antenna arrays and 802.11ac BWs using W-LS estimator.

6.6. Measurement Analysis for 2-D Wireless Indoor Positioning

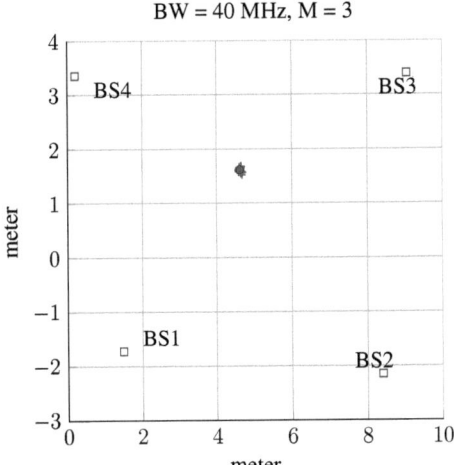

Figure 6.41: Localization results using 40 MHz BW and 3 antennas for each BS.

in cm for various number of snapshots based on the spatial diversity ($M = 1, \ldots, 6$) using 40 MHz BW. By investigating the CDFs of the position error, it has been noted that it is enough to increase the number of antennas for spatial diversity to 4 to achieve considerable improvement. It is worth mentioning that using only one iteration in the W-LS estimator achieves excellent performance. From the previous results of TDOA estimation using 1-D MP algorithms, it should be noted that using multiple snapshot principle based on the spatial diversity improves the performance considerably. Results have been introduced in [56].

6.6.2 Using 2-D MP Algorithms for 2-D Wireless Indoor Positioning

If the BSs of the wireless indoor positioning system are equipped with antenna arrays, the 2-D MP algorithms can be used to estimate not only the time delays, but also the relative DOAs. The hybrid TDOA and DOA estimators are used to estimate the MU coordinates. However, it is interesting first to investigate the performance of using TDOA and DOA observations individually, and then the hybrid TDOA and DOA estimators.

6.6.2.1 Position RMSE versus System Bandwidths and Number of Antennas using TDOA Estimates

In this subsection, the previous investigation of using 1-D UMP-Ex will be repeated here by using 2-D UMP. The TDOA observations have been used to estimate the MU position. Fig. 6.43 shows the RMSE of MU position estimation in cm versus 802.11ac BWs while three antennas have been used for each array in all four BSs. The performance of all MU coordinates estimators presented in Table 5.1 based on the TDOA observations has been presented. Chan estimator achieves the worst performance at low system BW (20 MHz

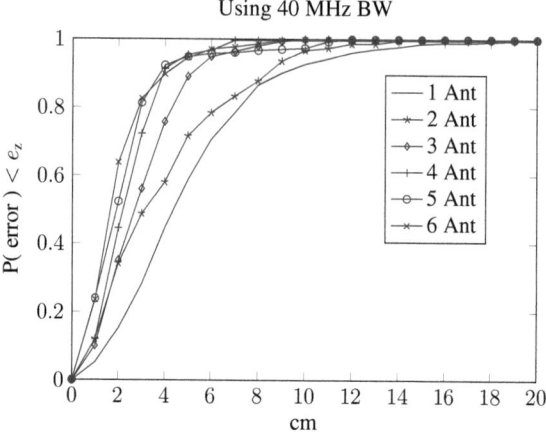

Figure 6.42: CDFs of the position error e_z in cm for various number of snapshots based on the spatial diversity using 40 MHz BW.

BW) while the RMSE is 14 cm. Fig. 6.44 shows the RMSE of MU position estimation in cm while the system BW is fixed at 20 MHz, and the number of antennas in all arrays has been changed equally between $M = 2$ and $M = 6$. From the above, it can be observed that the estimators with weights provide a modest improvement over non-weighted estimators, especially at low order of system parameters. The weighted ILS estimator is close to the W-LS in our experimental results, where the MU is located approximately in the center of the interested area. For another scenario, this estimator has a problem to converge, because it strongly depends on the initial value, which is taken usually the center of the interested area. The W-LS estimator is the best regarding the complexity and accuracy among the six considered estimators. Fig. 6.45 shows the performance of W-LS estimator versus all system BWs and antenna array orders.

The coming conclusion can be revealed from the above figures of TDOA scenario, if the system BW is limited, the performance can be improved by using the multiple snapshot principle based on the spatial diversity. For example from Fig. 6.45, by using 20 MHz BW with six antennas for each BS, the RMSE of W-LS is 5.66 cm, and by using 80 MHz BW with 2 antennas for each BS, the RMSE of W-LS is 4.27 cm.

6.6.2.2 Position RMSE versus System Bandwidths and Number of Antennas using DOA Estimates

In this section, the performance of using DOA observations resulting from using 2-D UMP to estimate the MU position will be investigated. Fig. 6.46 shows the RMSE in cm of the estimated MU position versus system BWs while $M = 4$, and Fig. 6.47 shows the RMSE versus antenna array order and 40 MHz BW. The performance of W-SLS estimator is the best estimator over the seven considered estimators, where the performance of LS estimator is the worst. The performance of LS and SLS estimators is improved

6.6. Measurement Analysis for 2-D Wireless Indoor Positioning

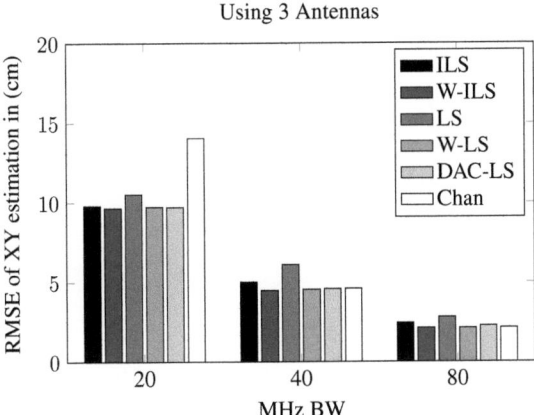

Figure 6.43: Comparison of XY coordinates estimation accuracy of various 802.11ac BWs and MU position estimators while $M = 3$, based on the TDOA observations.

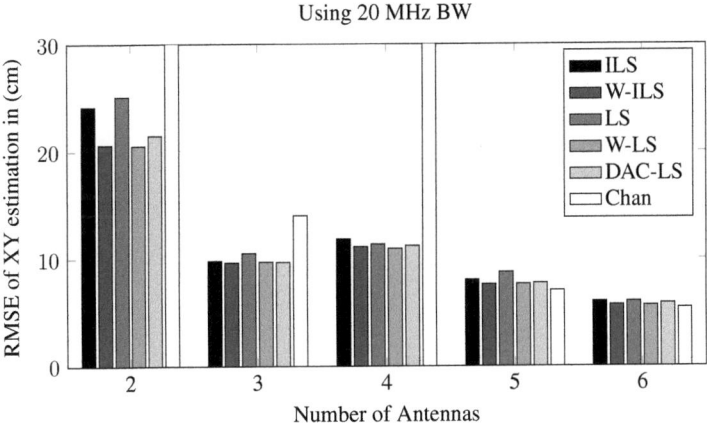

Figure 6.44: Comparison of XY coordinates estimation accuracy based on the TDOA observations for different number of array elements and 20 MHz BW.

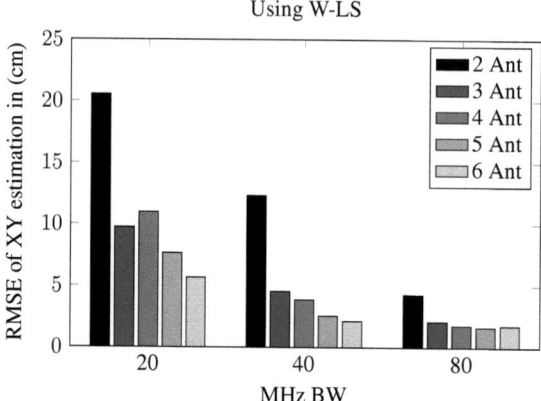

Figure 6.45: The performance of W-LS estimator based on TDOA observations versus system BWs and antenna array orders.

by including the weights, and the performance of DAC based on LI (DAC-LI) has been improved a little bit in A-DAC-LI by including all estimated DOAs of pairs, which are not located in the front of each other. It should also be noted that LI estimator achieves reliable accuracy, where all possible of BS pairs have been considered if the BSs of the discussed pair are not located in the front of each other. One thing also improves the accuracy of LI estimator is the results mitigation of noisy pairs if the corresponding two lines are not intersected (the estimated DOAs in the discussed two BSs are noisy). From the above two figures, if the order of antenna arrays is limited, the performance of using DOA observations can be improved by using wide BWs. For example, by using 40 MHz BW and $M = 6$, the RMSE of W-SLS is 7.37 cm as shown in Fig. 6.47. And by using 80 MHz BW and $M = 4$, the RMSE of W-SLS is 7.1 cm as shown in Fig. 6.46.

We now evaluate the distribution of the position error defined by (6.3) for all estimators presented in Table 5.1 that are based on DOA observations. Fig. 6.48, 6.49, and 6.50 show that for three cases: (M = 4, BW = 20 MHz), (M = 4, BW = 40 MHz), and (M = 4, BW = 80 MHz), respectively. Results reveal the following: the weighted estimators provide better performance than the non-weighted estimators especially if the BW increases. The gap between the performance of both types increases by increasing the system BW. In general, the performance of W-SLS and A-DAC-LI are the best. In case of A-DAC-LI, if the lines of the investigated pair are not intersected, there is no result to be included. It should be noted that the LI estimator does not include any weight, but the DAC-LI or A-DAC-LI, which are based on lines intersection, give a weight to each outcome of each discussed pair to get the final MU position estimate as described in Section 5.2.3.

6.6.2.3 Position RMSE using Hybrid TDOA and DOA

So far, the performance of using the outcomes of 2-D UMP algorithm (time delay and DOA) has been investigated individually. The following conclusions can be drawn from

6.6. Measurement Analysis for 2-D Wireless Indoor Positioning

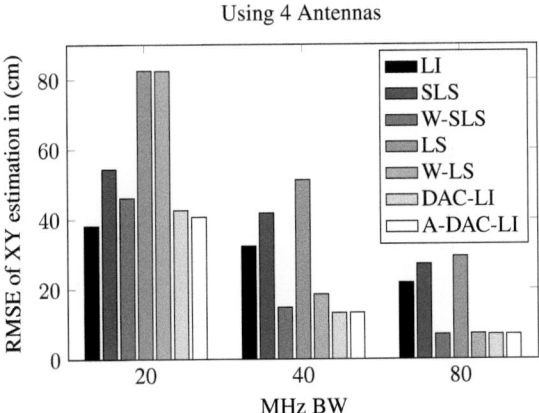

Figure 6.46: Comparison of XY coordinates estimation accuracy based on DOA observations of various 802.11ac BWs and MU position estimators while $M = 4$.

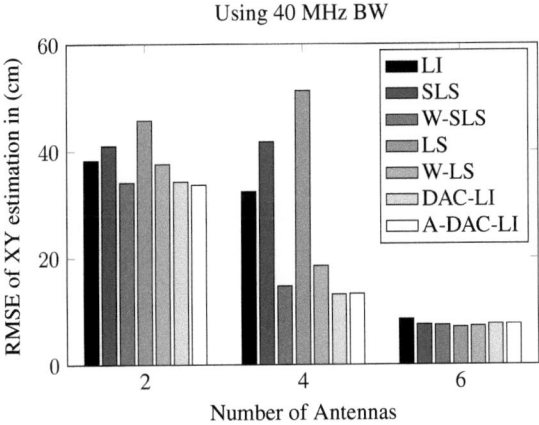

Figure 6.47: Comparison of XY coordinates estimation accuracy based on DOA observations for different number of array elements and 40 MHz BW.

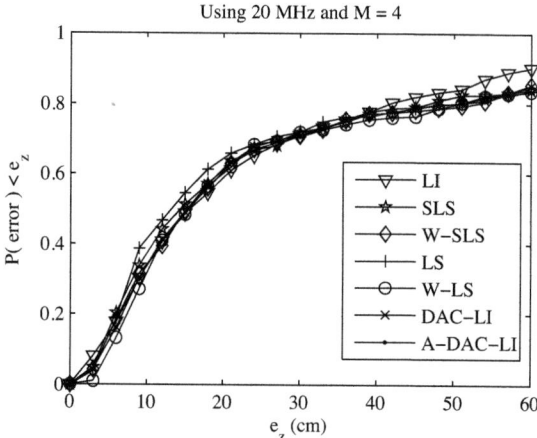

Figure 6.48: The distribution of the position error e_z of various estimators based on DOA observations (BW = 20 MHz, M = 4).

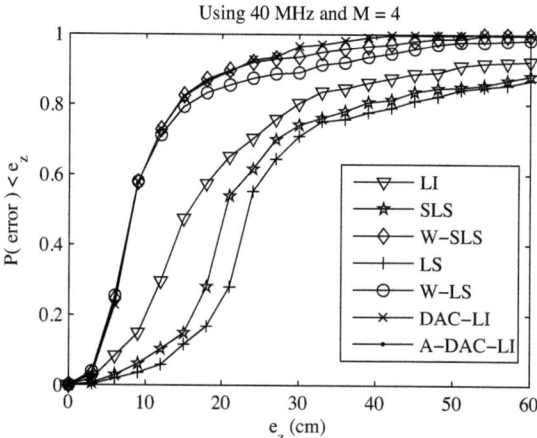

Figure 6.49: The distribution of the position error e_z of various estimators based on DOA observations (BW = 40 MHz, M = 4).

6.6. Measurement Analysis for 2-D Wireless Indoor Positioning

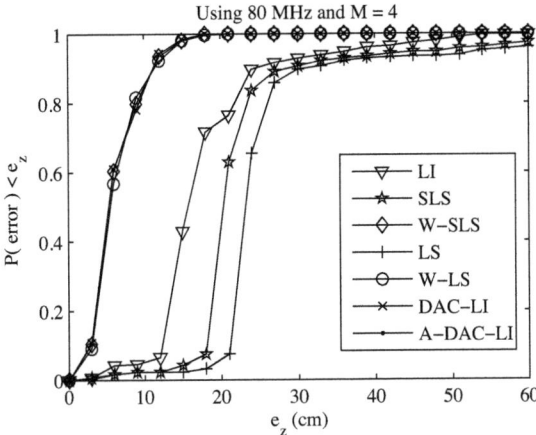

Figure 6.50: The distribution of the position error e_z of various estimators based on DOA observations (BW = 80 MHz, M = 4).

the previous results: the performance of the proposed W-LS estimator is the best for MU position estimation using the TDOA observations, and the performance of W-SLS estimator is the best for MU position estimation using the DOA observations. Let us now compare between the performance of these two estimators to estimate the MU position. Fig. 6.51 shows a comparison between the RMSE of W-LS estimator using TDOA observations, and W-SLS estimator using DOA observations. Results, shown for three cases $M = 2$, $M = 4$, and $M = 6$ with 20, 40, and 80 MHz BWs, reveal the following: using TDOA observations for localization is more accurate than using DOA observations although both of them have been extracted from the same subspace by using the same eigenvectors (refer to Section 4.3.2). As a result, to combine the results of using TDOA and DOA observations, appropriate weights should be given based on their range-based accuracy. Those weights cannot be constants, where each environment or each time of recording has its own characteristics. As it has been described in Chapter 5, the covariance matrix of TDOA observations and the variance matrix of DOA observations converted to distance form have been used. Those variance matrices are used to scale the different accuracy for the final result. To do that there are two options. The first option is to use them to scale the outcomes of both estimators to get the final estimate of MU position such as in Hybrid-DAC estimator presented in Section 5.3.1 and Hybrid-W estimator presented in Section 5.3.2. The second option is to use them to scale the TDOA and DOA observations inherently using one estimator such as the proposed Hybrid-WLS estimator presented in Section 5.4. It can also be observed that if the system BW is fixed, for example at 20 MHz, the achieved accuracy is not increased smoothly by increasing the number of antennas. Logically, the accuracy should be increased by increasing the number of antennas till by a very small amount. However, the number of snapshots is not that large (310 snapshots), and they were recorded during eight days and at different times during the day. Another reason behind that could be the limited accuracy of antenna arrays fabrication, where the distance between antenna elements is not exactly equal. In the following, the performance of using the above hybrid estimators will be presented.

6.6.2.3.1 Position RMSE using DAC based on TDOA and DOA Estimator
The performance of Hybrid-DAC estimator is presented in Fig. 6.52, where the RMSE is plotted versus system BW and number of antennas. It has been observed that the position RMSE is reduced by increasing the number of antennas and system BW considerably. It should also be noted that increasing system BW is more useful for MU localization than a number of antennas. The reason behind that is the accuracy of using TDOA is more than that of using DOA.

6.6.2.3.2 Position RMSE using Weighted TDOA and DOA Estimator
Similar to the previous paragraph where the impact of increasing M and system BW have been evaluated for Hybrid-DAC estimator. We now investigate the resulting impact of those parameters using Hybrid-W estimator. The outperformed W-LS among the other six estimators using TDOA observations, and the outperformed W-SLS among the other seven estimators using DOA observations are used. From the first look to Fig. 6.53, it can be observed that a modest performance improvement has been occurred compared to the previous estimator at low order of M and low system BW. However, the performance is comparable to the previous estimator at high order of M and system BW. Similar results are obtained by increasing the number of antennas and system BW.

6.6.2.3.3 Position RMSE using Proposed Hybrid TDOA and DOA Estimator
In this paragraph also, the impact of arrays order and system BW has been investigated for the proposed Hybrid-WLS estimator as shown in Fig. 6.54. It provides the excellent performance of MU positioning using joint time delay and DOA estimation among the three estimators at low system conditions. At low system conditions, a sensitive improvement has been occurred, and that is due to processing on both estimates of TDOA and DOA once. A considerable accuracy has been obtained at wide BWs and high order of antenna arrays. For example, if the system BW is 80 MHz and $M = 3$, the RMSE is 1.58 cm.

Finally, it is interesting also to show the error distributions of the three previous estimators. Two sceneries will be presented: low system conditions ($M = 2$, and BW = 20 MHz), and good system conditions ($M = 3$, and BW = 80 MHz) as shown in Fig. 6.55. Results, shown for two cases ($M = 2$, and BW = 20 MHz) and ($M = 3$, and BW = 80 MHz), reveal the following: the proposed hybrid estimator provides the excellent performance among the three considered estimators. The curve of its CDF is the sharpest among the three estimators. Furthermore, we can conclude that using TDOA and DOA observations in a single estimator and scale the different accuracy inherently achieves better performance than the problem dividing to two parts such as the principle of the hybrid estimator based on weighting or that of hybrid DAC, which divides the observations to some sets and some pairs. The performance of the proposed hybrid estimator can also be recognized from the constellation of the estimated XY coordinates on the XY plane as shown in Fig. 6.56 for the two mentioned cases of low system conditions and good system conditions.

6.6. Measurement Analysis for 2-D Wireless Indoor Positioning

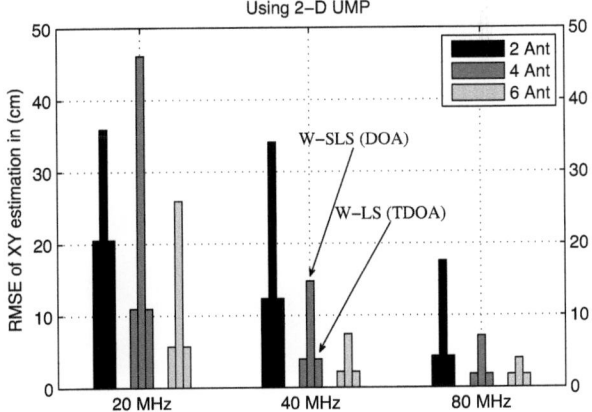

Figure 6.51: Comparison between the RMSE of W-LS estimator using TDOA observations, and W-SLS estimator using DOA observations.

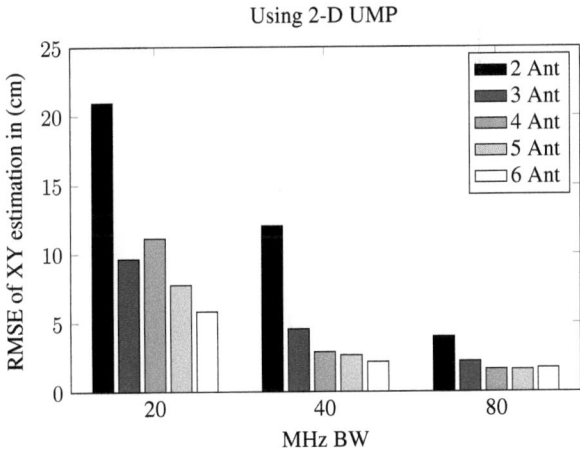

Figure 6.52: The performance of hybrid DAC estimator versus number of antennas and system BWs.

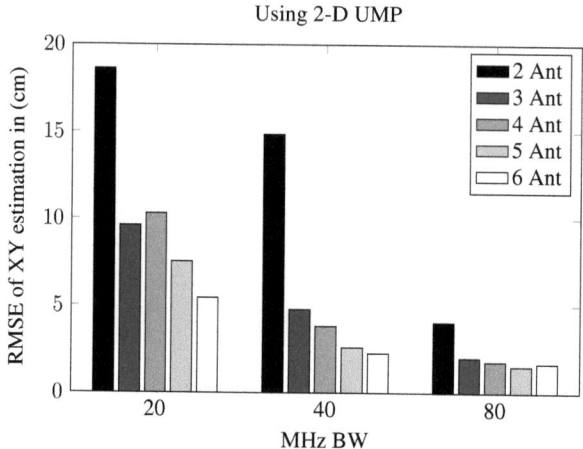

Figure 6.53: The performance of the weighted TDOA and DOA estimator versus number of antennas and system BWs.

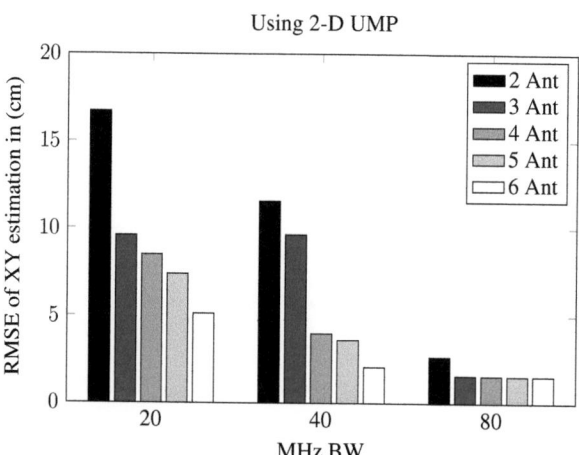

Figure 6.54: The performance of the proposed hybrid TDOA and DOA estimator versus number of antennas and system BWs.

6.6. Measurement Analysis for 2-D Wireless Indoor Positioning

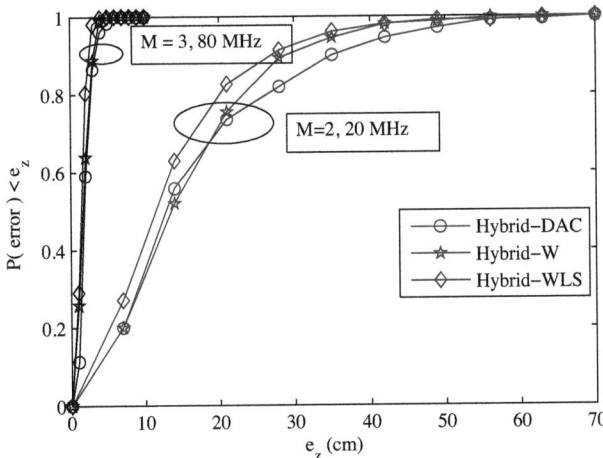

Figure 6.55: The distribution of the position error e_z of various hybrid estimators based on TDOA and DOA observations.

6.6.3 Performance Comparison between 1-D MP with Spatial Diversity and 2-D MP Algorithms

So far, the performance of 1-D MP using the effective 1-D UMP-Ex, and that of 2-D MP using the effective 2-D UMP has been presented. Let us now compare between their performances to show which is more robust for wireless indoor positioning if the BSs are perfectly synchronized. The result of 1-D MP processing is the time delays of the effective paths in the channel profile, and the results of 2-D MP processing are the time delays and the relative DOAs of the effective paths in the channel profile for all BSs within the service.

Let us now combine the results in Fig. 6.40 of 1-D UMP-Ex with the proposed W-LS estimator, and the results in Fig. 6.54 of 2-D UMP with the proposed hybrid TDOA and DOA estimator to get Fig. 6.57. It can be observed that at low system conditions, the accuracy of 1-D UMP-Ex with spatial diversity is exceeding that of 2-D UMP. However, if the antenna arrays order increases beyond 5, the 2-D UMP performs better than the 1-D UMP-Ex with the cost of complexity (ϱ will be smaller than 0.23).

Finally, it is interesting to confirm the previous conclusion by presenting the position error distributions through the CDFs. Let us fix the system BW to be 40 MHz, and show the CDFs of both techniques versus increasing number of CFRs measured from antenna array elements in each BS. They are used in a multiple snapshot principle for 1-D UMP-Ex and steering vectors for 2-D UMP. The number of antennas in each BS has been configured to be 2, 3, or 4. Results have been presented in Figs. 6.58, 6.59, 6.60, respectively. In all figures, the CDF of 1-D UMP-Ex is sharper than that of 2-D UMP. Consequence, it can be concluded that the accuracy of MU position estimation using 1-D MP

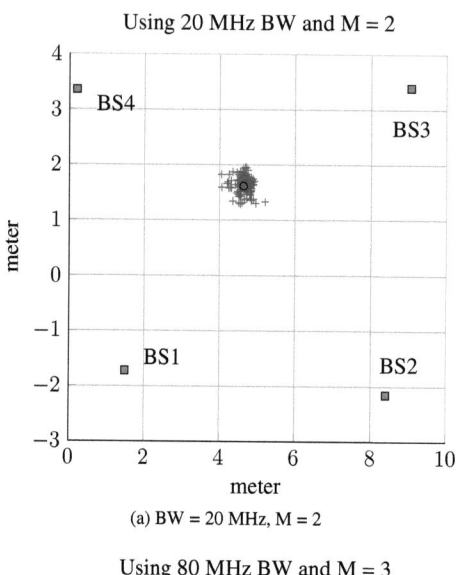

(a) BW = 20 MHz, M = 2

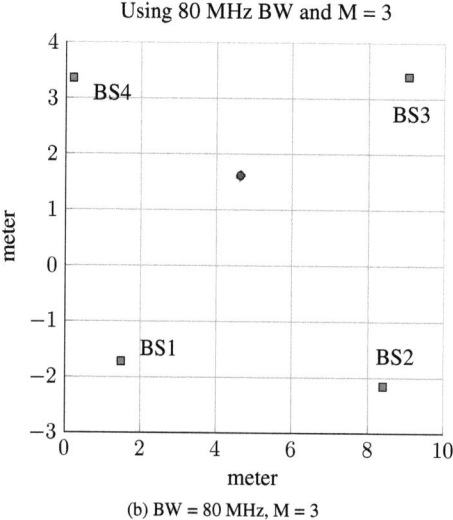

(b) BW = 80 MHz, M = 3

Figure 6.56: Localization results using (BW = 20 MHz, $M = 2$) and (BW = 80 MHz, $M = 3$)

6.6. Measurement Analysis for 2-D Wireless Indoor Positioning

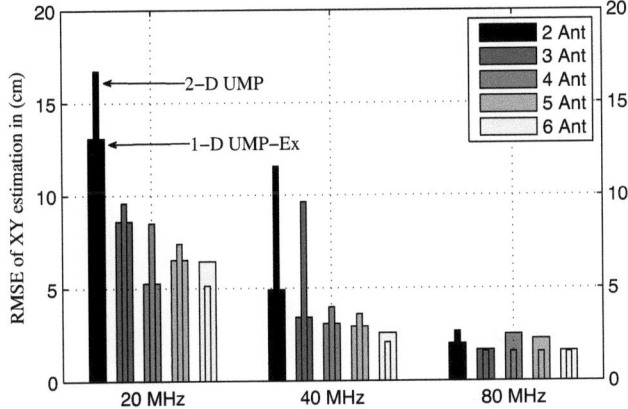

Figure 6.57: Performance comparison of 1-D UMP-Ex versus 2-D UMP.

techniques with spatial diversity is exceeding that of using 2-D MP techniques if the BSs are perfectly synchronized. The 1-D MP algorithms do not only succeed in the achieved accuracy, but also in the complexity, as it has been presented in Section 4.8 through Fig. 4.7. Really, it is surprising that our result at low system conditions reveals that joint time delay and DOA estimation by ULA-OFDM technology does not give an improvement on the position accuracy over than that obtained from time delay estimation using SIMO-OFDM technology while the antenna array is used for spatial diversity. In another way, we can say, using DOA observations for positioning gives a negligible improvement in the positioning accuracy over than that obtained from using TDOA observations if the BSs are perfectly synchronized. However, if the BSs cannot be synchronized perfectly, the hybrid TDOA and DOA should be outperformed than using only TDOA with spatial diversity by assuming that the RF daughter-boards of antenna elements of each BS can be synchronized perfectly. Table 6.3 shows a comparison between 1-D MP with spatial diversity and 2-D MP algorithms.

6.6.4 2-D Wireless Indoor Positioning at a Number of Static Positions

In this section, the performance of 1-D UMP-Ex and 2-D UMP to estimate the coordinates of the transmitter is presented. The transmitter is located at three static positions. The XY coordinates of the transmitter in the lab are at position 1 (313,160), position 2 (463,160), and position 3 (613,160) in cm, where the test environment presented in Fig. 6.35 is used. For each Rx antenna, 10 measurements were recorded for averaging purposes. The system BW is configured to be 40 MHz and the number of array elements is 4 for each BS. The other system parameters have been configured such as the previous. The proposed W-LS estimator is used for 1-D UMP-Ex, and the proposed hybrid estimator is used for 2-D UMP. Fig. 6.61 and 6.62 show the localization results using 1-D UMP-Ex and 2-D UMP, respectively.

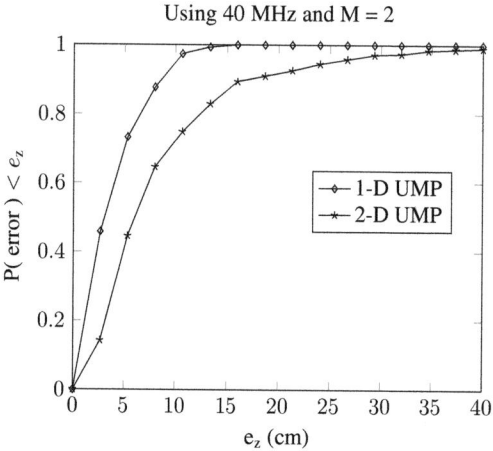

Figure 6.58: CDFs of the position error e_z while $M = 2$.

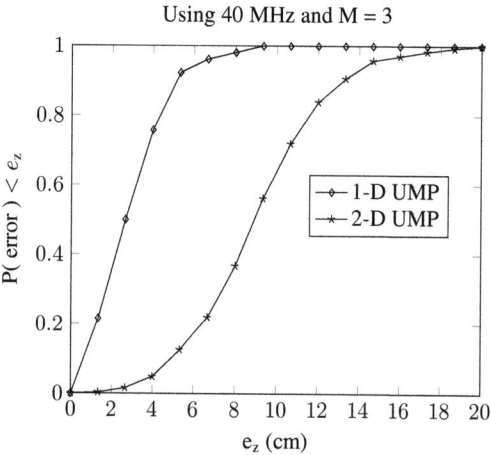

Figure 6.59: CDFs of the position error e_z while $M = 3$.

6.6. Measurement Analysis for 2-D Wireless Indoor Positioning 165

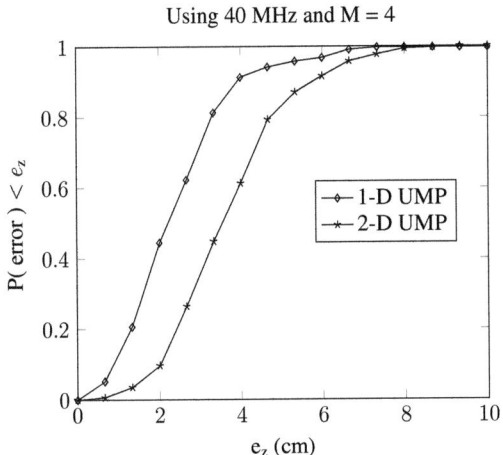

Figure 6.60: CDFs of the position error e_z while $M = 4$.

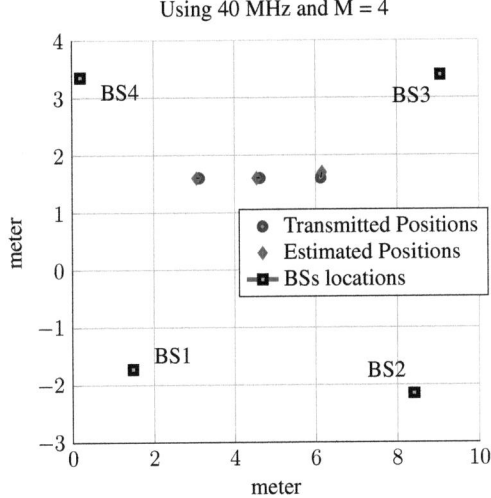

Figure 6.61: Localization results using 40 MHz BW and 4 antennas for each BS using 1-D UMP-Ex.

Table 6.3: Comparison between 1-D MP algorithms with spatial diversity and 2-D MP algorithms.

Description	1-D MP with Spatial Diversity	2-D MP with Antenna Arrays
Complexity	Less complex, the complexity ratio of 1-D UMP-Ex to 2-D UMP is 0.23	More complex, the complexity ratio of 2-D UMP to 1-D UMP-Ex is 4.3
Array imperfections	Antenna elements work as a kind of spatial diversity, hence, it is not sensitive to antenna array imperfections.	The space dimension depends on the steering vector accuracy, hence, it is sensitive to antenna array imperfections, hence, calibration should be made.
Accuracy	More accurate (especially at low system parameters) if the BSs are perfectly synchronized.	Less accurate (especially at low system parameters). It is more useful for non-perfectly synchronized BSs.
Arrays orientation	The orientation of all antenna arrays in the whole system should not be known.	The orientation of all antenna arrays in the whole system should be known.
MU position vs. array orientation	It does not depend on the arrival angle, see (5.125).	The DOA observation depends on the arrival angle, hence, the location of MU is important with respect to each antenna array in all BSs within the service, see (5.109).
Distance between MU and BS	It has no impact on the time delay estimation if the first path is still detected *.	Using DOA for positioning depends on the estimated value of θ, hence, if the MU is far away from the BS, it leads to high positioning error *.

* It is worth mentioning that if the Tx and Rx are in a NLOS, there is an induced propagation time delay, which depends on the material characteristics of the objects that are located in the media between them. However, the induced propagation time delay could be assumed as an insignificant parameter.

6.6. Measurement Analysis for 2-D Wireless Indoor Positioning 167

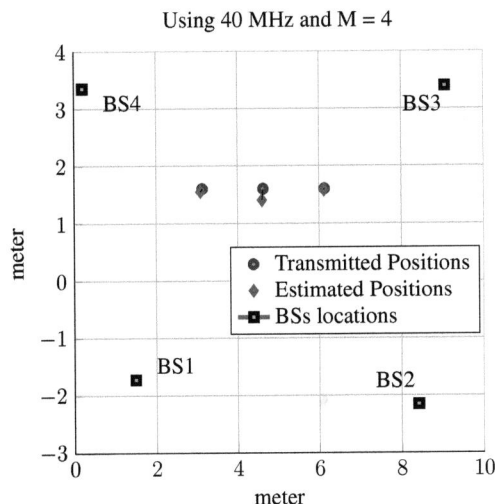

Figure 6.62: Localization results using 40 MHz BW and 4 antennas for each BS using 2-D UMP.

CHAPTER 7
UDP Identification for High-Resolution Wireless Indoor Positioning

In indoor environments, the MU is often in a NLOS state, and the direct path could be completely blocked due to the harsh nature of indoor environments. Therefore, the estimated time delay of the first path should be identified either as a very weak DDP (detected direct path) or even as an UDP (undetected direct path). Consequently, precise estimation of the channel profile parameters is not enough for high-resolution wireless indoor positioning system. However, it stays representing a key element to identify the UDP condition. In the previous, the accurate estimation of channel profile parameters has been addressed using the MP algorithms. In this chapter, the proper modeling of DDP and UDP channel profiles will be treated and addressed to the problem of UDP identification.

7.1 Introduction to the Problem of UDP Channel

From the previous, a precise wireless positioning system can be obtained by measuring the time delays of the received signals at a number of BSs. The accuracy of time-based wireless positioning systems depends strongly on the precise estimation of the FDP. The indoor environments are commonly known as dense multipath propagation environments with high probability of NLOS signal propagation. Hence, the estimated time delay of the FDP could be larger than the real time delay of the direct path between the MU and the BS. Consequently, the estimated distance is positively biased. Using the estimated time delays of the UDP channel profiles degrades the positioning accuracy significantly. Therefore, one of the major challenges for wireless indoor positioning is the identification of UDP condition. Adding the channel obstruction knowledge improves the accuracy of the positioning system. The results can be discarded or rectified if there is a limited connectivity.

Several methods have been proposed mainly for NLOS identification in the cellular networks and recently for the UWB technology. These methods can be classified into three groups [135]. The first one is based on range estimates. In [136], the variance of range measurements is used to identify the LOS channels in the cellular domain. It has been assumed that the MU is moving, the surrounding obstacles are varying, and the variance of LOS range measurements is known. Hence, the variance of NLOS range measurements is very large, and it changes over the time. The running variance-based methods have a latency. On the other hand, the MU is mostly static in wireless indoor networks. Hence, the deviation between the LOS and NLOS variances is negligible, and a wrong decision can be occurred. The second one is based on channel statistics. Here, one or more features from the received signal are extracted to identify the channel profile condition. The mean excess delay, the RMS delay spread, and the Kurtosis parameters have been used in [82]. The Kurtosis is a measure of how peaky a sample data, and it is defined as the ratio of the fourth moment of the data to the square of the second-order moment of the data (the variance) [82]. The Kurtosis provides information about the amplitude statistics of the multipath components. It has been found that the above statistics can be modeled by log-normal distribution; the IEEE 802.15.4a UWB channel models are utilized rather than that obtained from measurements. In [63], the RMS delay spread and the total power are used and modeled by normal and Weibull distributions, respectively. The channel profiles have been obtained from the measurements with a system BW of 500 MHz centered around 1 GHz. The channel profile parameters have been estimated using the conventional chirp-z method along with raised cosine filter. The third one is based on the electronic map of the building. The ray-tracing is used as in [137], where the electronic map including walls and other obstacles should be known. However, it takes a long time in large buildings, and it is difficult to incorporate each wall and all obstacles in the electronic map.

The objectives of this chapter are three-fold. First, the problem of estimating the propagation time delays and the relative amplitudes of multipath signals has been investigated using a single transmitter and a number of receiving BSs in both scenarios of DDP and UDP conditions. Second, the statistics of the observable channel profile parameters in both DDP and UDP conditions are presented using IEEE 802.11 wideband signals. Third, the performance of the proposed method for UDP identification is presented.

The rest of the chapter is organized as follows: system model and channel profile hypotheses are presented in Section 7.2. The statistics of channel profile parameters are presented in Section 7.3 through a number of experiments. The likelihood-ratio test is presented in Section 7.4. System performance and conclusions are presented in Section 7.5.

7.2 System Model

In time-based wireless indoor positioning, the time delay of the FDP is used as an estimate time delay of the direct path denoted by $\hat{\tau}_1$. The main features of $\hat{\tau}_1$ have been investigated in Section 2.5, which includes the main sources of the DME, $\xi_d = |\hat{d} - d|$, in time-based wireless indoor positioning. Based on Section 2.5, the estimated channel profiles can be classified mainly into two types (two hypotheses) based on the availability of the direct path: the DDP and the UDP channel profiles

$$\begin{aligned} S_0 &: DDP \quad \hat{\tau}_1 - \tau_1 \text{ is very small} \Rightarrow \xi_d \approx 0 \\ S_1 &: UDP \quad \hat{\tau}_1 - \tau_1 \text{ is very large} \Rightarrow \xi_d \gg 0 \end{aligned} \quad (7.1)$$

where S_0 is the DDP hypothesis, and S_1 is the UDP hypothesis. For DDP hypothesis, the DME is very small and the estimated parameters of the channel profile can be used to estimate the MU coordinates. For UDP hypothesis, the DME is very large and the estimated parameters of the channel profile should be mitigated.

7.3 Statistical Modeling of Multipath Channel Features

The estimated parameters of the channel profile represent the key element to identify the channel condition. To investigate that let us describe first the measurement system.

Experimental results are presented now to show the capability of using the enhanced 1-D UMP-Ex for a high-resolution estimate of the propagation time delays and the channel gains for wireless indoor positioning with UDP identification capability. The following system environments are used to investigate the statistics of DDP and UDP conditions as shown in Fig. 7.1. For DDP investigation, four BSs are located in the lab in LOS with the transmitter while the other two BSs are located in the corridor in NLOS with the transmitter (the direct path is still detected). The wall has a thickness of 13 cm. The location of the transmitter has been switched between three positions in the middle of the lab. For UDP investigation, two BSs are located in the antenna room in NLOS with the transmitter, where the direct path is corrupted by the metallic chamber, and two BSs are located in the corridor in NLOS with the transmitter, where the direct path is corrupted by the elevator shaft and most of the time by the elevator itself. The location of the transmitter has been switched between three positions as shown in Fig. 7.1. The height of all antennas in the system was 150 cm. The omni-directional antennas with 3 dBi gain were used. The network analyzer, Agilent ENA E5071C, was used to measure the indoor CFR in the frequency range of 802.11ac. A carrier frequency of 5.25 GHz was used. The transmitted power of the ENA was 10 dBm. The time delays of the connecting cables have been measured using the wideband TDT, Agilent 86100A.

The complex CFR of each antenna element can be obtained by sweeping the channel at uniformly spaced frequencies. The real and imaginary parts of the forward transmission coefficient S21 were measured and stored for further processing. For averaging purposes and accurate statistics, 9000 measurements were recorded for DDP condition during ten days. And 6000 measurements were recorded for UDP condition during seven days. Each measurement covers 160 MHz BW with a sampling interval set to Δf =312.5 KHz. The number of samples of the CFR is 512 samples, which is equal to the maximum IFFT/FFT order of 802.11ac.

Let us now investigate the time delay and power characteristics of the received radio signals in both DDP and UDP conditions. A single snapshot from a LOS channel and a NLOS channel corrupted by a chamber have been used. Fig. 7.2 shows the estimated time delays and the relative amplitudes of the effective paths in both scenarios. The following observations can be made: (a) In DDP condition, the FDP is the strongest path. However, that is not necessary, because in most NLOS environments, the direct path is not the strongest path, but it can be detected. (b) In UDP condition, the FDP is usually not the strongest path, and the amplitudes of the successive paths are comparable to the first path. Also the received signals are highly attenuated compared to that of DDP condition. (c) The mean time delay of the UDP channel profiles is larger than that of the DDP channel profiles. Therefore, the total power and the mean time delay of the received signal can be used to identify the UDP condition. To do that the statistics of multipath channel profiles in both scenarios should be investigated.

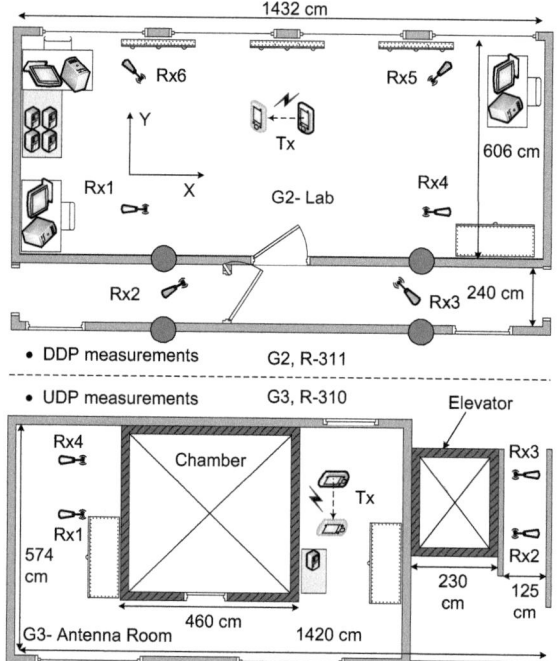

Figure 7.1: The test environment.

7.3. Statistical Modeling of Multipath Channel Features

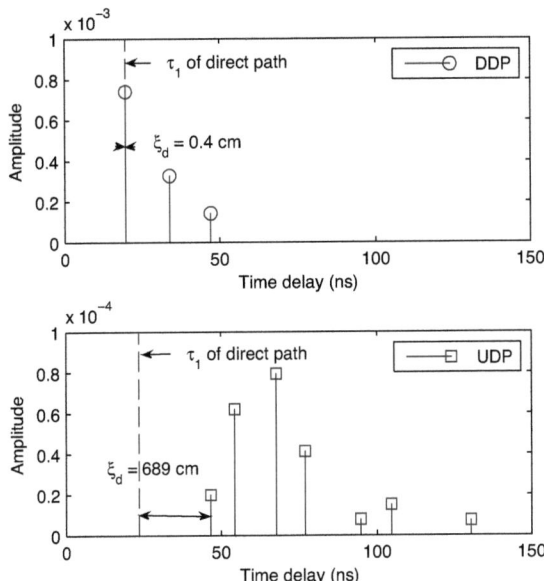

Figure 7.2: The estimated time delays and the relative amplitudes of typical DDP and UDP channel profiles using 160 MHz BW.

7.3.1 Mean Excess Time Delay

The mean excess time delay τ_{MED} of the multipath channel is an important parameter that can be calculated from the estimated parameters of the $\hat{L}$ effective paths as in (2.8)

$$\tau_{MED} = \sum_{l=1}^{\hat{L}} \hat{\tau}_l |\alpha_l|^2 / \sum_{l=1}^{\hat{L}} |\alpha_l|^2,$$

where $|\alpha_l|^2$ is the power of lth path. From the above equation and Fig. 7.2, it can be found that the parameter τ_{MED} of the DDP channel profiles should be smaller than that of the UDP channel profiles.

In order to determine the goodness-of-fit of τ_{MED} in both DDP and UDP conditions, the Anderson-Darling (AD) statistic, it is also known as empirical cumulative distribution function (ECDF), is used to measure how well the given data fit the 16 common different distributions using Minitab statistical tool. The distribution of the smallest adjusted AD statistic value offers the best fit; the data points follow the straight line well. The confidence level of this test has been selected to be 95%. It has been found that τ_{MED} can be best modeled with the normal distribution. The probability plots and the PDFs of τ_{MED} in DDP and UDP conditions are shown in Fig. 7.3. To show the difference more obviously, the estimated τ_{MED} has been converted to distance by multiplication by the speed of light, c. Clearly, an observable gap between DDP and UDP statistics is available, hence, the normal distribution parameters in both scenarios are distinctive. The normal distribution of both scenarios can then be described as

$$p(\tau_{MED}/S_0) = \frac{1}{\sqrt{2\pi}\sigma_0} e^{-(\tau_{MED}-\mu_0)^2/2\sigma_0^2} \qquad (7.2)$$

$$p(\tau_{MED}/S_1) = \frac{1}{\sqrt{2\pi}\sigma_1} e^{-(\tau_{MED}-\mu_1)^2/2\sigma_1^2} \qquad (7.3)$$

where μ_0 and σ_0 are the mean and standard deviation of the normal distribution of the DDP channel profiles. Similarly, μ_1 and σ_1 are the mean and standard deviation of the normal distribution of the UDP channel profiles. Table 7.1 presents the statistics of τ_{MED} in both scenarios.

7.3.2 Total Power

The total power of the received signal is defined as

$$P_{tot} = 10 \log_{10}(\sum_{l=1}^{\hat{L}} |\alpha_l|^2). \qquad (7.4)$$

From (7.4) and Fig. 7.2, it is worth mentioning that the total power of the UDP channel profiles should be more attenuated than that of the DDP channel profiles.

Such as the previous, the AD statistic is used to determine the best fit. To show the difference more clearly between the total power of UDP and DDP channel profiles, let us present the results in terms of power loss instead of total power as, $P_{loss} = -P_{tot}$. It has been found that the P_{loss} parameter can also be best modeled with the normal distribution. Fig. 7.4 shows the probability plots and the PDFs of the P_{loss} parameter in both DDP and UDP conditions. The observable gap between DDP and UDP power loss statistics allows

7.3. Statistical Modeling of Multipath Channel Features 175

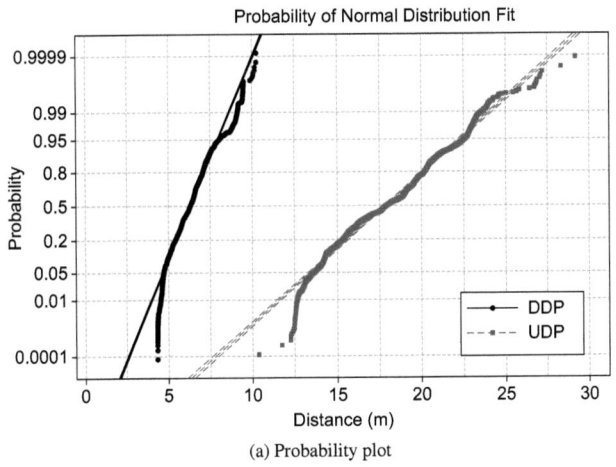

(a) Probability plot

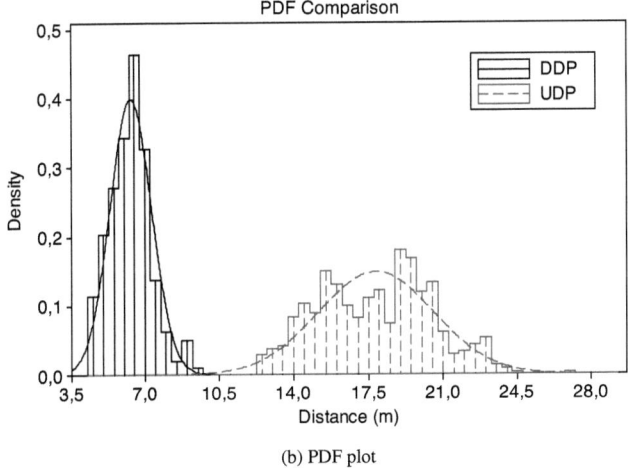

(b) PDF plot

Figure 7.3: Validity of modeling τ_{MED} by normal distribution.

us to distinguish between both conditions. The normal distribution of both conditions can be described as in (7.2) and (7.3) for DDP and UDP conditions, respectively, in terms of the P_{loss} parameter. The mean and standard deviation of the P_{loss} parameter in both scenarios are presented in Table 7.1. It is worth mentioning that the gap between the PDF of UDP and that of DDP in case of the P_{loss} parameter is larger than that in case of the τ_{MED} parameter, hence, using P_{loss} for UDP identification should lead to better results.

7.3.3 Hybrid Time-Power Parameter

Besides τ_{MED} and P_{loss} parameters, a hybrid parameter can be extracted from both parameters for UDP identification. Since the time delay and the relative amplitude of each effective path have been estimated from the measured CFR, the hybrid parameter can then be defined as

$$\kappa = \tau_{MED} \times P_{loss}. \quad (7.5)$$

From the previous, the statistics of the hybrid parameter κ in DDP conditions should be very smaller than that in UDP conditions. By using the AD statistic to determine the best fit, it has been found that the parameter κ can be best modeled with the log-normal distribution. Fig. 7.5 shows the probability plots and the PDFs of the parameter κ in both conditions. The observable gap between DDP and UDP statistics of the hybrid parameter κ allows us to distinguish between both conditions. The log-normal distribution can be described as

$$p(\kappa/S_i) = \frac{1}{\kappa\sqrt{2\pi}\sigma_i} e^{-(\ln(\kappa)-\mu_i)^2/2\sigma_i^2} \quad (7.6)$$

where the superscript $i = \{0, 1\}$ denotes the channel condition, μ_i and σ_i are the mean and standard deviation of $\ln(\kappa)$, respectively. The statistics of the κ parameter are presented in Table 7.1. It is worth mentioning that the time delay and the relative amplitude of the FDP only have been used in [63] to define a hybrid parameter, which could be valid for wireless indoor positioning based on the TOA observations. However, the time delays and the relative amplitudes of all effective paths should be used as in (7.5) for wireless indoor positioning based on the TDOA observations. It is worth mentioning that the BW of 160 MHz is used in the previous statistical figures.

7.4 Likelihood-ratio Test for UDP Channel Profile Identification

The normal PDFs of τ_{MED} and P_{loss}, and the log-normal PDF of κ in the DDP and UDP conditions are $p_{DDP}(\tau_{MED})$, $p_{UDP}(\tau_{MED})$, $p_{DDP}(P_{loss})$, $p_{UDP}(P_{loss})$, $p_{DDP}(\kappa)$, and $p_{UDP}(\kappa)$, respectively. If the statistics of those parameters are already known for DDP and UDP conditions using the results in Table 7.1, the binary likelihood-ratio test can be performed to select the more likely hypothesis [82]. To identify the channel profile condition, the following likelihood-ratio tests are used:

1. Mean Excess Time Delay Test:

$$\frac{p_{DDP}(\tau_{MED})}{p_{UDP}(\tau_{MED})} \underset{S_1}{\overset{S_0}{\gtrless}} 1 \quad (7.7)$$

7.4. Likelihood-ratio Test for UDP Channel Profile Identification

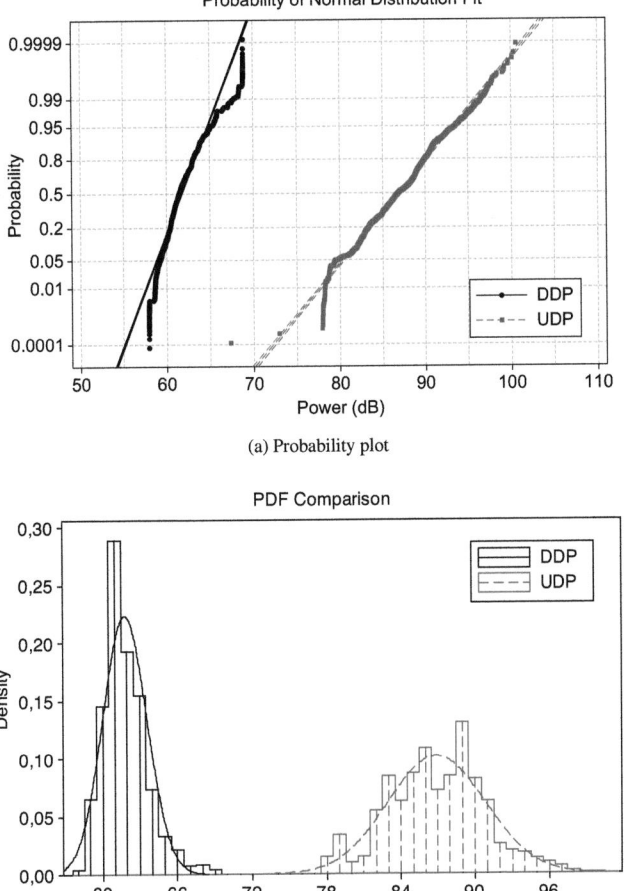

(a) Probability plot

(b) PDF plot

Figure 7.4: Validity of modeling P_{loss} by normal distribution.

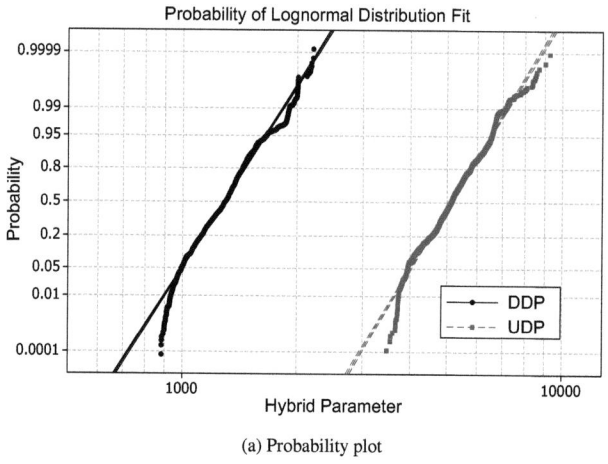

(a) Probability plot

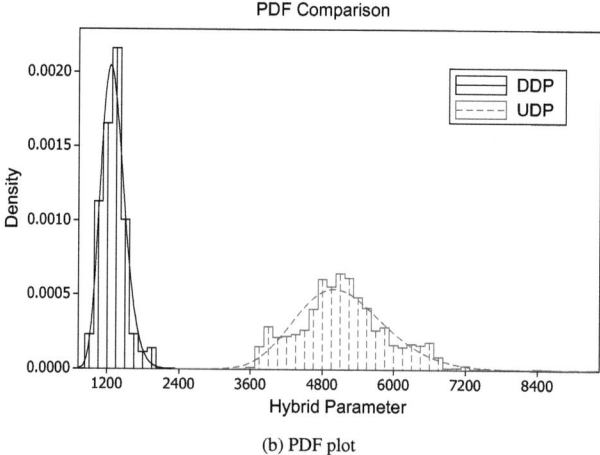

(b) PDF plot

Figure 7.5: Validity of modeling κ by log-normal distribution.

7.5. System Performance

2. Power Loss Test:
$$\frac{p_{DDP}(P_{loss})}{p_{UDP}(P_{loss})} \underset{s_1}{\overset{s_0}{\gtrless}} 1 \qquad (7.8)$$

3. Hybrid Time-Power Test:
$$\frac{p_{DDP}(\kappa)}{p_{UDP}(\kappa)} \underset{s_1}{\overset{s_0}{\gtrless}} 1 \qquad (7.9)$$

The outputs of the above likelihood-ratio tests are compared to one. If it is larger than one, the estimated channel profile is more likely to be of the DDP condition, and can be used to calculate the MU coordinates. Otherwise, the estimated channel profile is more likely to be of the UDP condition, hence, the estimated time delay of the first path should be discarded. Only the remaining BSs, which have channel profiles belonging to the DDP condition should be used to estimate the MU coordinates. If the connectivity is limited, $\hat{\tau}_1$ of UDP channel can be rectified; it is still a future work.

From (7.7) to (7.9), the PDF of each parameter has been used individually to make a decision. However, a joint likelihood-ratio test can be formed by combining the likelihood tests of all parameters. A sub-optimal likelihood test can be obtained by assuming that the investigated parameters τ_{MED}, P_{loss}, and κ are independent, similar to the principle used in [82]. The sub-optimal likelihood function can then be defined as

$$\frac{p_{DDP}(\tau_{MED})}{p_{UDP}(\tau_{MED})} \times \frac{p_{DDP}(P_{loss})}{p_{UDP}(P_{loss})} \times \frac{p_{DDP}(\kappa)}{p_{UDP}(\kappa)} \underset{s_1}{\overset{s_0}{\gtrless}} 1. \qquad (7.10)$$

The decision of (7.10) should provide the highest accuracy for channel profile type detection.

7.5 System Performance

To show the performance of the proposed methods: the enhanced 1-D UMP-Ex algorithm for the propagation time delays and the relative amplitudes of multipath signals estimation, and the likelihood-ratio test for UDP channel profile identification, let us investigate the probability of UDP identification using 6000 channel profiles. The probability of UDP identification by using 20, 40, 80, and 160 MHz BWs of 802.11ac is shown in Fig. 7.6. A huge performance has been achieved due to the high-resolution estimation of the channel profile parameters. For example, the probability of UDP identification is more than 96.6% at the smallest BW (20 MHz). It is better than the reported results in the survey of [135], although the UWB technology is used in most of the proposed methods. The highest probability of UDP identification using channel measurements was 89.29% presented in [81] while the system BW was 500 MHz around 1 GHz. Results have been introduced in [59].

To show the validation of the proposed algorithms in different environment using the previous system setup, let us record the indoor CFRs of another harsh environment such as in Fig. 7.7. It shows the stairs of building 2 and 3 in the university of Magdeburg. Fig. 7.8 shows the inside view of the measurement environment between floors 3 and 2. The transmitter position has been switched between seven positions in the third floor as shown in Fig. 7.7. The position of the receiver has also been switched between seven positions

Table 7.1: The mean and standard deviation of the normal PDFs for τ_{MED} and P_{loss}, and the log-normal PDF for $\ln(\kappa)$ using 802.11ac parameters.

Ch. Parameter \ BW MHz			20	40	80	160
τ_{MED} (ns)	DDP	μ_0	25.91	25.46	22.73	21.05
		σ_0	7.14	6.52	3.75	3.33
	UDP	μ_1	65.95	61.98	60.86	59.47
		σ_1	28.66	10.96	9.35	8.98
P_{loss} (dB)	DDP	μ_0	61.12	61	61.34	61.8
		σ_0	2.95	2.74	2.19	1.79
	UDP	μ_1	86.87	86.61	86.27	86.87
		σ_1	4.66	3.61	3.74	3.93
$\ln(\kappa)$	DDP	μ_0	7.33	7.32	7.23	7.16
		σ_0	0.25	0.23	0.16	0.15
	UDP	μ_1	8.60	8.57	8.55	8.54
		σ_1	0.30	0.18	0.15	0.14

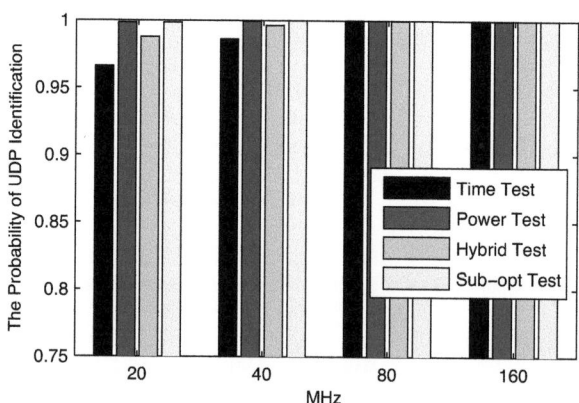

Figure 7.6: Probability of UDP identification using the proposed methods and 802.11ac signals.

7.5. System Performance

in the second and first floor. For each position, 250 measurements have been recorded. The Concrete thickness of each floor is around 38 cm.

Although a rigid obstruction is available between the transmitter and the receiver, it has been found that the receiver located in the second floor (one floor down) can detect approximately the direct path; the RMSEs of various 802.11ac BWs (20, 40, 80, and 160 MHz) are 1.59, 0.95, 0.46, and 0.28 meter, respectively. However, the receiver located in the first floor (two floors down) cannot detect the direct path; the RMSEs of various 802.11ac BWs (20, 40, 80, and 160 MHz) are 9.54, 4.12, 2.88, and 3.03 meter, respectively. From the RMSEs (the DMEs), it is worth mentioning that the direct path between the transmitter and the receiver located in the first floor is totally blocked. However, the RMSEs or the DMEs of using 80 or 160 MHz BWs are smaller than the expected. As a result, the first path that has been estimated comes through the concrete objects of the two floors, and not through the free space between stairs. Hence, let us now check the ability of the proposed algorithms for UDP correct identification using the estimated channel parameters of this receiver. The statistical results of Table 7.1 are used. Fig. 7.9 shows the probability of UDP correct identification. Such as the previous, an excellent performance has been obtained.

As a conclusion, the 1-D UMP-Ex algorithm is used to estimate the time delays and the relative amplitudes for a high-resolution wireless indoor positioning system. The channel profile statistics and the proper modeling of both DDP and UDP conditions using 802.11ac signals have been presented for UDP identification. Experimental results using the emerging IEEE 802.11ac standard reveal that the achieved probability of correct identification can be more than 96.6% at the smallest bandwidth. The high accuracy of UDP identification leads to precise positioning. As a future work, channel models should be developed for wireless indoor positioning taking into account different indoor areas. Appropriate weights could be given to the results of UDP channels if the connectivity is limited for calculating the MU coordinates. The results of 2-D UMP are the amplitudes, the time delays, and the DOAs of the multipath channel components, hence, the DOAs can also be used to identify the channel condition besides the investigated parameters of the channel profile.

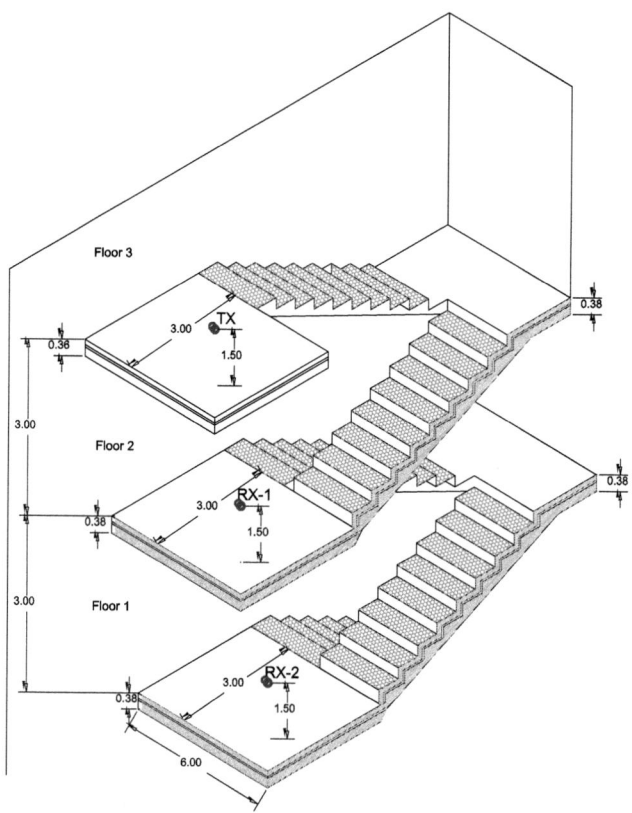

Figure 7.7: Another test environment (the stairs of building 2 and 3 in the university of Magdeburg).

7.5. System Performance

Figure 7.8: The real environment between floor 3 and floor 2 of building 2 and 3 in the university of Magdeburg.

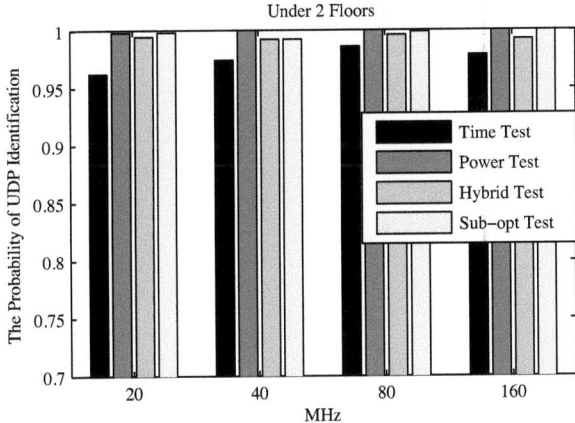

Figure 7.9: Probability of UDP identification using the proposed methods and 802.11ac signals, where the receiver is located under two floors.

CHAPTER 8
Conclusions and Future Work

8.1 Conclusions

In the following, we briefly summarize the conclusions drawn from our research work presented in this dissertation. More details can be found in the previous chapters.

In the future, the LBS based on the wireless indoor positioning systems will be widely used, which need accurate position estimation in indoor environments. The traditional positioning systems such as GPS or cellular positioning cannot provide that in indoor environments. Therefore, there is a need for new positioning systems that are designed for indoor applications and can overcome the challenges of indoor environments. As a result, the wireless indoor positioning is emerging as a new important research area. In this work, an effective wireless indoor positioning has been developed based on the opportune signals of IEEE 802.11 standards. The emerging 802.11ac standard for the 5 GHz band enhances the MIMO order to 8×8 and system BW to 160 MHz compared to 802.11n and 802.11a. The performance of using wideband orthogonal multi-carrier signals and multi antennas for wireless indoor positioning has been presented.

The fundamentals of position estimation have been presented such as the wireless positioning architecture, the model of indoor radio propagation channel, the possible locations of MU inside the building, and many others. The characteristics of radio signals including RSS, DOA, TOA, and TDOA have been presented with their advantages and disadvantages. Since our concern is the high accuracy, the time-based wireless positioning techniques have been preferred in this work. One from our goals is that the MU should be a very simple device, and the complexity could be added only to the network side, therefore, the TDOA principle has been preferred in this work. In fact, using TDOA principle for wireless positioning can operate with system elements using their normal frame formats. In addition, the main sources of DME in time-based wireless indoor positioning systems have been presented.

The channel profile parameters such as the time delays and the relative amplitudes of a multipath channel can be extracted from the CFR. Therefore, the principle of CFR and CIR of the multipath wireless channel has been presented. To measure the CFR, the preamble of the OFDM frame has been used. The LTS of 802.11a and the LTF of 802.11n and 802.11ac have been preferred to achieve the highest range of estimated time delays. The discontinuity of the estimated CFR is solved by doing an interpolation, hence, the time delay estimation problem of OFDM signal in a multipath channel is fully equivalent to the DOA problem using the ULA in antenna array processing. In fact, the CFR of the multipath radio channel can be seen as a harmonic signal model. Therefore, the super-resolution algorithms of DOA estimation can be enhanced for OFDM time delay estimation.

High-accuracy with low-complexity position estimation is essential for many applications using recent wireless networks. Recently, variants of MP algorithms have been presented to estimate the DOA of coherent or non-coherent signals using ULA. They are working with data directly without forming a covariance matrix, and the snapshot-by-snapshot analysis is used. The DOA in the presence of multipath coherent signals can be estimated without performing additional processing such as the other subspace-based algorithms.

In this work, the problem of highly resolving the propagation time delays and the relative amplitudes associated with signals in a multipath communication channel is addressed for wireless positioning based on the TDOA observations. The recent subspace-based algorithms, represented by MP, UMP, and BMP algorithms, are applied in a new way to estimate those parameters from the measured CFR using OFDM systems. In fact, the UMP and BMP algorithms are enhanced from the MP algorithm by doing a real computation to reduce the complexity, especially that of SVD computation. To evaluate the performance of various MP algorithms more precisely, the empirical CFR should be employed. Therefore, a number of experiments have been made in typical indoor environments.

The accuracy, stability, and complexity of the forward MP, and UMP, the forward-backward MP-Ex, and UMP-Ex, SBMP, and MBMP algorithms are investigated for TDOA estimation using 802.11ac parameters, and compared to the corresponding performance of 802.11a and 802.11n. The performance of using wideband orthogonal multi-carrier signals and diversity techniques are presented. Using the wide BWs of 802.11ac improves the accuracy and stability of TDOA estimation, especially at low SNR. Using multiple LTFs per OFDM frame in multiple snapshot principle increases the accuracy and stability also. However, using both of them increases the complexity of TDOA estimation. To reduce the complexity of high BWs of 802.11ac, it can be treated in multiple snapshot principle based on the frequency diversity. To reduce the complexity of the 160 MHz BW, it can be treated as two snapshots of 80 MHz BW. Accuracy and STD in the range of a few hundreds of picoseconds are achieved using various MP algorithms. The MBMP and UMP-Ex provide better performance than the others, and they have the same accuracy when all invariances of MBMP are used. The complexity ratio of UMP-Ex to MBMP is 0.70. Therefore, UMP-Ex is the best choice for TDOA estimation applications.

Since the most-recent wireless generations use the MIMO-OFDM technology such as 802.11 ac, the availability of multiple antennas in BSs can be used to estimate the DOA as another observation and as a kind of spatial diversity for TDOA estimation besides the frequency diversity coming from OFDM. Therefore, the hybrid TDOA and DOA observations have been used. If the ULA is used with OFDM, then we have an array of two dimensions, the first dimension is the OFDM pilots, distributed equally in the frequency dimension, and the second dimension is the array elements, distributed equally in the space dimension. It represents the key element for us to enhance the 2-D MP algorithms, which have been presented based on the uniform space between array elements using the URA to estimate multiple frequencies or azimuth and elevation angles. In this work, the 2-D MP, 2-D UMP, and 2-D BMP algorithms have been implemented in different realizations to estimate the propagation time delays and the relative DOAs for a high-resolution wireless positioning system. Those recent subspace-based algorithms are applied in a new way to estimate these parameters simultaneously from the measured S-CFR using multiple antennas and wideband orthogonal multi-carrier signals. Using the priori information of wireless positioning (our concern is to estimate the time delay and the DOA of the

8.1. Conclusions

shortest path) mitigates the problem of repeated poles and hence reduces the complexity of calculating extra EVD problems.

From the results of LOS and NLOS experiments, using multi-antenna multi-carrier principles can successfully enhance the dimensionality of the signal subspace for joint time delay and DOA estimation; it represents a robust technique versus multipath channel fading. The performance of the 2-D MP, 2-D MP-Ex, 2-D UMP, and the 2-D BMP algorithms have been investigated and compared using 802.11ac system parameters. The 2-D UMP is the best choice for wireless positioning regarding the complexity and accuracy. Based on the NLOS experimental results, accuracy in the range of a few hundreds of picoseconds, and a fraction of one degree could be achieved.

The necessary number of antennas for accurate DOA estimation can be reduced by increasing the number of subcarriers (BW), which reduces the receiver complexity. In general, if the order of arrays is limited, the performance can be improved by using wide BWs, and if the system BW is limited, the performance can be improved by using arrays of high order. It is worth mentioning that the diversity and the improvement techniques provide a significant enhancement to the performance if the system BW is limited, but if the system BW is wide enough, they provide only small enhancement. It has been observed that the position RMSE is reduced by increasing the number of antennas and system BW. However, increasing system BW is more useful for MU localization than increasing the number of antennas. The reason behind that is the accuracy of using TDOA is more than that of using DOA.

The principle of 1-D MP algorithms are also enhanced to estimate the propagation time delays from the S-CFR using multi-antenna multi-carrier systems for a high-resolution wireless positioning. Using the spatial diversity principle in 1-D MP algorithms improves the accuracy of wireless positioning considerably especially at low SNR and narrow system BWs; it represents a robust technique versus multipath channel fading.

From the previous, in addition to spectral diversity coming from OFDM, two principles are presented to use the spatial diversity through 1-D and 2-D MP algorithms. The required complexity for TDOA estimation with spatial diversity and that for hybrid TDOA and DOA estimation is compared. The complexity ratio of 1-D UMP-Ex to 2-D UMP is around $\varrho = 0.23$ at all BWs of 802.11ac. Our experimental results using the emerging 802.11ac standard reveal that using DOA observations besides TDOA observations for positioning in 2-D UMP gives a negligible improvement to the positioning accuracy over than that obtained from using TDOA observations if the BSs are perfectly synchronized. However, if the BSs cannot be synchronized perfectly, the hybrid TDOA and DOA should be outperformed than using only TDOA with spatial diversity by assuming that the RF daughter-boards of antenna elements of each BS could be synchronized and calibrated perfectly. At low system conditions, the accuracy of 1-D UMP-Ex with spatial diversity is exceeding that of 2-D UMP. However, if the antenna arrays order increases beyond 5, the 2-D UMP performs better than the 1-D UMP-Ex with the cost of complexity (ϱ will be smaller than 0.23). Using the spatial diversity principle in 1-D UMP-Ex reduces the complexity and increases the accuracy. The reasons behind that are: it mitigates the problem of antenna arrays imperfections and orientations that are typical in DOA. The accuracy of DOA estimates depends on the MU position with respect to the antenna array of each BS within the service. The distance between the MU and the BS has a negligible impact on the time delay estimation if the direct path is still detected while in 2-D UMP if the MU is far away from the BS, the DOA estimates lead to high positioning error.

To estimate the MU coordinates, the non-iterative W-LS estimator is presented based on the TDOA estimates, and the non-iterative hybrid W-LS estimator is also presented based on the TDOA and DOA estimates. The TDOA and DOA observations have been given some weights due to the different accuracy of each type. The proposed estimators solve the problem of the initial guess and that of partitioning of the other estimators. They are robust against channel fading and low SNR. The lower bounds, represented by the CRB, of TDOA and DOA estimation error variances are derived in terms of the fundamental parameters of the wireless indoor positioning system such as the antenna array order, number of subcarriers including the subcarrier spacing, SNR, and the DOA.

Experimental results using 802.11ac system parameters and the proposed techniques show that accuracy in the range of 1.5 cm can be achieved for 2-D wireless indoor positioning. The location of BSs inside the building should be designed properly. The achieved accuracy is very useful not only for positioning but also for communication applications.

Since the MU is often in a NLOS state, the direct path could be completely blocked due to the harsh nature of indoor environments. Therefore, the estimated time delay of the first path should be identified either as a very weak DDP or even as a UDP. The channel profile statistics and the proper modeling of both DDP and UDP conditions using 802.11ac signals have been treated and addressed for UDP identification. The mean excess time delay and the power loss have been used. The PDF of mean excess time delay is best modeled by normal distribution; similarly, the power loss is modeled. The mean excess time delay and the power loss are multiplied to get the hybrid parameter, which is best modeled by log-normal distribution. The binary likelihood ratio test has been used to identify the channel profile condition. Experimental results using the emerging 802.11ac standard reveal that the achieved probability of correct identification can be more than 96.6% at 20 MHz BW. If the estimated channel profile is more likely to be of the UDP condition, the estimated time delay of the shortest path should be discarded. Only the remaining BSs, which have channel profiles belonging to the DDP condition should be used to estimate the MU coordinates. The results can be rectified if there is a limited connectivity. Adding the channel obstruction knowledge improves the accuracy of the positioning system.

8.2 Future Work

As it has been stated, the wireless indoor positioning is a new research field having multiple research aspects. There are many research topics that should be more investigated such as channel measurement and modeling, develop new positioning algorithms, solve the synchronization problems based on the capability of the commercial oscillators, and many others. In fact, the applications of wireless indoor positioning systems are different, as a result, the required accuracy and the system requirements are different. Those different scenarios make the research in the area of indoor positioning interesting and challenging.

The following specific projects can be investigated as a continuation of the research work in this dissertation that are based on the TDOA and DOA estimation techniques. In this work, it has been preferred to use the WLAN for wireless indoor positioning, where

8.2. Future Work

the multipath channel parameters such as the amplitudes, the relative time delays, and the relative DOAs have been estimated. The estimated parameters can be used for joint communication and positioning. They can be used for equalization assistance and beam forming especially in the downlink.

In this work, the ULA is used in the BSs, however, different antenna arrays can be used such as the URA. The various 2-D MP algorithms can then be easily extended to 3-D MP algorithms to do joint estimation for time delays, and both azimuth and elevation angles. Another extension of this work can be the utilization of the spatial diversity in the MU transmission, since most of MU devices are equipped with two antennas. A few BSs are occurring with circular antenna array, therefore, the 2-D MP algorithms should be enhanced to work with those kind of BSs.

The CRBs of TDOA and DOA estimation error variances for a single path channel, and that of joint time delay and DOA estimation error variance for a multipath channel have been derived to investigate the lower bound of their error variances. It has been assumed that the time delay and the DOA are unbiased parameters. In practice, the DME is biased by NLOS propagation channel. Therefore, the CRB of TDOA and DOA including the NLOS effects can be derived for MIMO-OFDM systems. In this work, the variances of TDOA and DOA observations have been used to give a weight to each observation. Another extension of this work is to investigate the capability of obtaining the appropriate weights from the estimated channel profile parameters without using the statistical variances.

In this work, a limited number of UDP channel profiles have been investigated. Therefore, channel models should be developed for wireless indoor positioning taking into account different indoor areas. In case of UDP condition, if the connectivity is limited for calculating the MU coordinates, appropriate weights could be given to the results of UDP channels. To identify the UDP condition, the amplitudes and the relative time delays have been used. However, the results of 2-D MP algorithms are the amplitudes, the time delays, and the DOAs of the multipath channel components, hence, the DOAs can also be used to identify the channel condition besides the investigated parameters of the channel profile.

On the other hand, tracking and positioning of multiple objects need to be investigated. The DME can be reduced adaptively by developing some algorithms for tracking. The electronic map of the building can also be used to correct the final position estimate.

Appendices

APPENDIX A
The Unitary Matrix Transformation

A square matrix $\mathbf{B}_{N \times N}$ is called unitary if it satisfies

$$\mathbf{B}^{-1} = \mathbf{B}^H. \tag{A.1}$$

Therefore, $\mathbf{B}^H \mathbf{B} = \mathbf{I}$, where $\mathbf{I}$ is the identity matrix. As a result, the columns of the matrix $\mathbf{B}$ must be orthonormal.

A matrix $\mathbf{A}_{M \times P}$ is called a centro-hermitian matrix [138] if it satisfies

$$\mathbf{A} = \underset{M}{\mathbf{\Pi}} \mathbf{A}^* \underset{P}{\mathbf{\Pi}} \tag{A.2}$$

where $\underset{q}{\mathbf{\Pi}}$ is called the exchange matrix, which is a square matrix $(q \times q)$. The components of the exchange matrix are zeros except for ones on the anti-diagonal as

$$\underset{q}{\mathbf{\Pi}} = \begin{bmatrix} 0 & \cdots & 1 \\ \vdots & \ddots & \vdots \\ 1 & \cdots & 0 \end{bmatrix}. \tag{A.3}$$

The exchange matrix $\underset{q}{\mathbf{\Pi}}$ reverses the ordering of the rows.

The UMP algorithm uses the following three theorems, where their proofs can be found in [90], [138] and many other references therein.

Theorem 1: If a vector $\mathbf{x} = [x_0, x_1, \cdots, x_{N-1}]^T$ is centro-hermitian, where N is an odd number, consequence, $\underset{N}{\mathbf{\Pi}} \mathbf{x}^* = \mathbf{x}$, then the matrix $\mathbf{Y}$

$$\mathbf{Y} = \begin{bmatrix} x_0 & x_1 & \cdots & x_P \\ x_1 & x_2 & \cdots & x_{P+1} \\ \vdots & \vdots & \ddots & \vdots \\ x_{N-P-1} & x_{N-P} & \cdots & x_{N-1} \end{bmatrix}_{(N-P) \times (P+1)} \tag{A.4}$$

is also centro-hermitian, where each column of $\mathbf{Y}$ is a windowed part of $\mathbf{x}$.

Theorem 2: The following matrix $[\mathbf{Y} : \underset{N-P}{\mathbf{\Pi}} \mathbf{Y}^* \underset{P+1}{\mathbf{\Pi}}]$ is always a centro-hermitian matrix, where $\mathbf{Y}$ is any complex $(N-P) \times (P+1)$ matrix.

Theorem 3: If a matrix $\mathbf{Y}$ of size $(M \times P)$ is centro-hermitian, then $\mathbf{Q}_M^H \mathbf{Y} \mathbf{Q}_P$ is a real matrix, where $\mathbf{Q}_M$ and $\mathbf{Q}_P$ are unitary matrices whose columns are conjugate symmetric and have a sparse structure as in (3.37) and (3.38) for even and odd order, respectively, [94].

APPENDIX B
Kronecker and Khatri-Rao Products

The Kronecker product is often used to build up repeated copies of small matrices. Those matrices are often zeros and ones. If $\mathbf{A}$ is an $N \times M$ matrix and $\mathbf{B}$ is a $K \times L$ matrix, then the Kronecker product $\mathbf{A} \otimes \mathbf{B}$ is the $NK \times ML$ matrix as the following [139], [116]

$$\mathbf{A} \otimes \mathbf{B} = \begin{bmatrix} a_{11}\mathbf{B} & \cdots & a_{1M}\mathbf{B} \\ \vdots & \ddots & \vdots \\ a_{N1}\mathbf{B} & \cdots & a_{NM}\mathbf{B} \end{bmatrix}. \tag{B.1}$$

The Khatri-Rao product of an $N \times M$ matrix $\mathbf{A}$ and a $P \times M$ matrix $\mathbf{B}$ is defined as the $NP \times M$ matrix [116]

$$\mathbf{A} \odot \mathbf{B} = [\mathbf{a}_1 \otimes \mathbf{b}_1 : \mathbf{a}_2 \otimes \mathbf{b}_2 : \cdots : \mathbf{a}_M \otimes \mathbf{b}_M] \tag{B.2}$$

where $\mathbf{a}_i$ and $\mathbf{b}_i$ denote the ith column of the matrices $\mathbf{A}$ and $\mathbf{B}$, respectively. The first submatrix of (B.2) is

$$\mathbf{a}_1 \otimes \mathbf{b}_1 = \begin{bmatrix} a_{11}\mathbf{b}_1 \\ \vdots \\ a_{N1}\mathbf{b}_1 \end{bmatrix}. \tag{B.3}$$

The interesting property between both products is

$$(\mathbf{A} \otimes \mathbf{B})(\mathbf{C} \odot \mathbf{D}) = \mathbf{AC} \odot \mathbf{BD}. \tag{B.4}$$

APPENDIX C
The QR and QZ algorithms

The generalized eigenvalue problem is defined:

$$(\mathbf{A} - \lambda\mathbf{B})\mathbf{q} = 0 \qquad (C.1)$$

where $\mathbf{A}$ and $\mathbf{B}$ are square matrices of size $L \times L$. The decomposition problem of matrix pencil pair $(\mathbf{A}, \mathbf{B})$ in (C.1) is reduced to a generalized Schur problem by QZ algorithm [83], [140]. The standard eigenvalue problem is defined:

$$(\mathbf{C} - \lambda\mathbf{I})\mathbf{q} = 0. \qquad (C.2)$$

From (C.1) and (C.2), it is clear that if $\mathbf{B} = \mathbf{I}$, the generalized eigenvalue problem reduces to a standard eigenvalue problem. The decomposition problem of (C.2) is reduced to a Schur form for the standard eigenvalue problem by QR algorithm [83], [140]. However, (C.2) can be obtained from (C.1) by using the Moore-Penrose pseudo-inverse of (3.26) as

$$\mathbf{C} = \mathbf{B}^\dagger \mathbf{A} = \left(\mathbf{B}^H \mathbf{B}\right)^{-1} \mathbf{B}^H \mathbf{A}. \qquad (C.3)$$

Bibliography

[1] M. Oziewicz, "On application of music algorithm to time delay estimation in ofdm channels," *Broadcasting, IEEE Transactions on*, vol. 51, no. 2, pp. 249–255, 2005.

[2] E. D. Kaplan, *Understanding GPS: Principles and Applications.* Artech House, 2006.

[3] W. J. ECKER, *Loran-C User handbook.* US Department of TransportatUS, United States Coast Guard, COMMANDANT PUBLICATION P16562.5, 1992.

[4] X. Li and K. Pahlavan, "Super-resolution toa estimation with diversity for indoor geolocation," *Wireless Communications, IEEE Transactions on*, vol. 3, no. 1, pp. 224–234, 2004.

[5] X. Li, "Super-resolution toa estimation with diversity techniques for indoor geolocation applications," Ph.D. dissertation, Worcester Polytechnic Institute, 2003.

[6] A. Bensky, *Wireless Positioning Technologies and Applications.* ARTECH HOUSE, 2008.

[7] V. A. Kushki A., Plataniotis K., "Indoor positioning with wireless local area networks (wlan)," in *Encyclopedia of GIS*, 2008, pp. 566–571.

[8] K. Kaemarungsi and P. Krishnamurthy, "Properties of indoor received signal strength for wlan location fingerprinting," in *Mobile and Ubiquitous Systems: Networking and Services, 2004. MOBIQUITOUS 2004. The First Annual International Conference on*, 2004, pp. 14–23.

[9] K. El-Kafrawy, M. Youssef, A. El-Keyi, and A. Naguib, "Propagation modeling for accurate indoor wlan rss-based localization," in *Vehicular Technology Conference Fall (VTC 2010-Fall), 2010 IEEE 72nd*, 2010, pp. 1–5.

[10] G. Athanaasiadou, A. Nix, and J. McGeehan, "A new 3d indoor ray-tracing propagation model with particular reference to the prediction of power and delay spread," in *Personal, Indoor and Mobile Radio Communications, 1995. PIMRC'95. Wireless: Merging onto the Information Superhighway., Sixth IEEE International Symposium on*, vol. 3, 1995, pp. 1161–.

[11] M. Lawton and J. McGeehan, "The application of a deterministic ray launching algorithm for the prediction of radio channel characteristics in small-cell environments," *Vehicular Technology, IEEE Transactions on*, vol. 43, no. 4, pp. 955–969, 1994.

[12] S. C. Spinella, A. Iera, and A. Molinaro, "On potentials and limitations of a hybrid wlan-rfid indoor positioning technique," *International Journal of Navigation and Observation*, vol. 2010, p. 11, 2010.

[13] S. Schwalowsky, H. Trsek, R. Exel, and N. Kero, "System integration of an ieee 802.11 based tdoa localization system," in *Precision Clock Synchronization for Measurement Control and Communication (ISPCS), 2010 International IEEE Symposium on*, 2010, pp. 55–60.

[14] M. Heidari, F. Akgul, and K. Pahlavan, "Identification of the absence of direct path in indoor localization systems," in *Personal, Indoor and Mobile Radio Communications, 2007. PIMRC 2007. IEEE 18th International Symposium on*, 2007, pp. 1–6.

[15] K. Pahlavan, F. Akgul, M. Heidari, A. Hatami, J. Elwell, and R. Tingley, "Indoor geolocation in the absence of direct path," *Wireless Communications, IEEE*, vol. 13, no. 6, pp. 50–58, 2006.

[16] S. Koo, C. Rosenberg, H.-H. Chan, and Y. C. Lee, "Location-based e-campus web services: from design to deployment," in *Pervasive Computing and Communications, 2003. (PerCom 2003). Proceedings of the First IEEE International Conference on*, 2003, pp. 207–215.

[17] "http://www.ettus.com/home."

[18] K. Pahlavan, X. Li, and J.-P. Makela, "Indoor geolocation science and technology," *Communications Magazine, IEEE*, vol. 40, no. 2, pp. 112–118, 2002.

[19] P. Jensfelt, "Approaches to mobile robot localization in indoor environments," Ph.D. dissertation, Royal Institute of Technology, Stockholm, 2001.

[20] L. Jing, P. Liang, C. Maoyong, and S. Nongliang, "Super-resolution time of arrival estimation for indoor geolocation based on ieee 802.11 a/g," in *Intelligent Control and Automation, 2008. WCICA 2008. 7th World Congress on*, 2008, pp. 6612–6615.

[21] K. Pahlavan and A. H. Levesque, *Wireless Information Networks*. NEW York: John Wiley and Sons, 2005.

[22] C. Patterson, R. Muntz, and C. Pancake, "Challenges in location-aware computing," *IEEE Pervasive Computing*, vol. 2, pp. 80–89, 2003.

[23] S. Ivanov, E. Nett, and S. Schemmer, "Automatic wlan localization for industrial automation," in *Factory Communication Systems, 2008. WFCS 2008. IEEE International Workshop on*, 2008, pp. 93–96.

[24] V. Prabhu and D. Jalihal, "An improved esprit based time-of-arrival estimation algorithm for vehicular ofdm systems," in *Vehicular Technology Conference, 2009. VTC Spring 2009. IEEE 69th*, April 2009, pp. 1–4.

[25] F. Zhao, W. Yao, C. Logothetis, and Y. Song, "Comparison of super-resolution algorithms for toa estimation in indoor ieee 802.11 wireless lans," in *Wireless Communications, Networking and Mobile Computing, 2006. WiCOM 2006.International Conference on*, 2006, pp. 1–5.

[26] M. Oziewicz, "The phasor representation of the ofdm signal in the sfn networks," *Broadcasting, IEEE Transactions on*, vol. 50, no. 1, pp. 63–70, 2004.

[27] M. Vanderveen, A.-J. van der Veen, and A. Paulraj, "Estimation of multipath parameters in wireless communications," *Signal Processing, IEEE Transactions on*, vol. 46, no. 3, pp. 682–690, 1998.

[28] R. Roy and T. Kailath, "Esprit-estimation of signal parameters via rotational invariance techniques," *Acoustics, Speech and Signal Processing, IEEE Transactions on*, vol. 37, no. 7, pp. 984–995, 1989.

[29] M. Zoltowski, M. Haardt, and C. P. Mathews, "Closed-form 2-d angle estimation with rectangular arrays in element space or beamspace via unitary esprit," *Signal Processing, IEEE Transactions on*, vol. 44, no. 2, pp. 316–328, 1996.

[30] M. Pesavento, A. Gershman, and M. Haardt, "Unitary root-music with a real-valued eigendecomposition: a theoretical and experimental performance study," *Signal Processing, IEEE Transactions on*, vol. 48, no. 5, pp. 1306–1314, May 2000.

[31] A.-J. van der Veen, M. Vanderveen, and A. Paulraj, "Joint angle and delay estimation using shift-invariance techniques," *Signal Processing, IEEE Transactions on*, vol. 46, no. 2, pp. 405–418, 1998.

[32] F. Ji, J. Liang, and F.-J. Chen, "Esprit algorithm for joint delay and 2-dimensional doa estimation of multipath parameters," in *Microwave, Antenna, Propagation and EMC Technologies for Wireless Communications, 2007 International Symposium on*, 2007, pp. 1093–1096.

[33] L. Cong and W. Zhuang, "Hybrid tdoa/aoa mobile user location for wideband cdma cellular systems," *Wireless Communications, IEEE Transactions on*, vol. 1, no. 3, pp. 439–447, 2002.

[34] Y. Hua and T. Sarkar, "Matrix pencil method for estimating parameters of exponentially damped/undamped sinusoids in noise," *Acoustics, Speech and Signal Processing, IEEE Transactions on*, vol. 38, no. 5, pp. 814–824, 1990.

[35] T. Sarkar and O. Pereira, "Using the matrix pencil method to estimate the parameters of a sum of complex exponentials," *Antennas and Propagation Magazine, IEEE*, vol. 37, no. 1, pp. 48–55, 1995.

[36] Y. Hua, "Estimating two-dimensional frequencies by matrix enhancement and matrix pencil," *Signal Processing, IEEE Transactions on*, vol. 40, no. 9, pp. 2267–2280, 1992.

[37] N. Yilmazer and T. Sarkar, "2-d unitary matrix pencil method for efficient direction of arrival estimation," *Digital Signal Processing*, vol. 16, no. 6, p. 767781, November 2006.

[38] M. F. Khan and M. Tufail, "Computationally efficient 2d beamspace matrix pencil method for direction of arrival estimation," *Digital Signal Processing*, vol. 20, no. 6, p. 15261534, December 2010.

[39] C. Liu, H. Jiang, and D. Wang, "Range-based node localization algorithm for wireless sensor network using unitary matrix pencil," in *Test and Measurement, 2009. ICTM '09. International Conference on*, vol. 2, Dec 2009, pp. 128–132.

[40] R. Ding, Z. Qian, and X. Wang, "Joint toa and doa estimation of ir-uwb system based on matrix pencil," in *Information Technology and Applications, 2009. IFITA '09. International Forum on*, vol. 1, 2009, pp. 544–547.

[41] T. J. S. Khanzada, A. Ali, and A. Omar, "Time difference of arrival estimation using super resolution algorithms to minimize distance measurement error for indoor positioning systems," in *Multitopic Conference, 2008. INMIC 2008. IEEE International*, Dec 2008, pp. 443–447.

[42] D. Torrieri, "Statistical theory of passive location systems," *Aerospace and Electronic Systems, IEEE Transactions on*, vol. AES-20, no. 2, pp. 183–198, 1984.

[43] Y. Chan and K. Ho, "A simple and efficient estimator for hyperbolic location," *Signal Processing, IEEE Transactions on*, vol. 42, no. 8, pp. 1905–1915, 1994.

[44] A. Urruela and J. Riba, "Novel closed-form ml position estimator for hyperbolic location," in *Acoustics, Speech, and Signal Processing, 2004. Proceedings. (ICASSP '04). IEEE International Conference on*, vol. 2, 2004, pp. 149–152.

[45] J. Abel, "A divide and conquer approach to least-squares estimation," *Aerospace and Electronic Systems, IEEE Transactions on*, vol. 26, no. 2, pp. 423–427, 1990.

[46] A. Urruela, A. Pages-Zamora, and J. Riba, "Divide-and-conquer based closed-form position estimation for aoa and tdoa measurements," in *IEEE International Conference on Acoustics, Speech and Signal Processing (ICASSP) 2006 Proceedings.*, vol. 4, May 2006, pp. IV921–924.

[47] Y. Oshman and P. , "Optimization of observer trajectories for bearings-only target localization," *and , IEEE Transactions on*, vol. 35, no. 3, pp. 892–902, 1999.

[48] *IEEE P802.11ac. Specification framework for TGac. IEEE 802.11- 09/0992r21*, Jan. 2011, Std.

[49] *Part 11: Wireless LAN medium access control (MAC) and physical layer (PHY) specifications, amendment 5: Enhancements for higher throughput*, IEEE Std 802.11-2009 Std.

[50] *Wireless LAN Medium Access Control (MAC) and Physical Layer (PHY) Specifications*, IEEE Std 802.11-2007 Std.

[51] T. S. Rappaport, *Wireless Communications Principles and Practice*. Prentice-Hall, 1996.

[52] A. Gaber and A. Omar, "Fundamentals of position estimation techniques," in *WMD: Wireless Positioning and Tracking in Indoor and Urban Environments: Methods, Architectures and Applications, IEEE International Microwave Symposium (IMS 2012), Montreal, CA*, 2012.

[53] ——, "Sub-nanosecond accuracy of tdoa estimation using matrix pencil algorithms and ieee 802.11," in *Wireless Communication Systems (ISWCS), 2012 International Symposium on*, 2012, pp. 646–650.

[54] ——, "A study of tdoa estimation using matrix pencil algorithms and ieee 802.11ac," in *Ubiquitous Positioning, Indoor Navigation, and Location Based Service (UPINLBS)*, Oct. 2012, pp. 1–8.

[55] ——, "Nlos wireless indoor positioning based on tdoa estimation using matrix pencil algorithms and ieee 802.11ac," *Journal of Global Positioning Systems*, 2013 (invited paper).

[56] ——, "Recent results of high-resolution wireless indoor positioning based on ieee 802.11ac," in *Radio and Wireless Symposium (RWS), 2014 IEEE*, Jan 2014, pp. 142–144.

[57] ——, "Joint time delay and doa estimation using 2-d matrix pencil algorithms and ieee 802.11ac," in *Positioning Navigation and Communication (WPNC), 2013 10th Workshop on*, March 2013, pp. 1–6.

[58] ——, "A study of wireless indoor positioning based on joint tdoa and doa estimation using 2-d matrix pencil algorithms and ieee 802.11ac," *Wireless Communications, IEEE Transactions on*, 2015.

[59] A. Gaber, A. Alsaih and A. Omar, "Udp identification for high-resolution wireless indoor positioning based on ieee 802.11ac," in *Positioning Navigation and Communication (WPNC), 2014 11th Workshop on*, March 2014.

[60] W. Yiding, W. Yirong, and H. Jun, "Application of inverse chirp-z transform in wideband radar," in *Geoscience and Remote Sensing Symposium, 2001. IGARSS '01. IEEE 2001 International*, vol. 4, 2001, pp. 1617–1619 vol.4.

[61] R. Mersereau, "An algorithm for performing an inverse chirp z-transform," *Acoustics, Speech and Signal Processing, IEEE Transactions on*, vol. 22, no. 5, pp. 387–388, Oct 1974.

[62] T. S. Rappaport, *Wireless Communications Principles and Practice*. Prentice Hall, 2002.

[63] M. Heidari, N. Alsindi, and K. Pahlavan, "Udp identification and error mitigation in toa-based indoor localization systems using neural network architecture," *Wireless Communications, IEEE Transactions on*, vol. 8, no. 7, pp. 3597–3607, 2009.

[64] M. Heidari and K. Pahlavan, "A model for dynamic behavior of ranging errors in toa-based indoor geolocation systems," in *Vehicular Technology Conference, 2006. VTC-2006 Fall. 2006 IEEE 64th*, 2006, pp. 1–5.

[65] D. Fittipaldi and M. Luise, "Cramer-rao bound for doa estimation with antenna arrays and uwb-ofdm signals for pan applications," in *Personal, Indoor and Mobile Radio Communications, 2008. PIMRC 2008. IEEE 19th International Symposium on*, 2008, pp. 1–5.

[66] M. Navarro and M. Najar, "Frequency domain joint toa and doa estimation in ir-uwb," *Wireless Communications, IEEE Transactions on*, vol. 10, no. 10, pp. 1–11, 2011.

[67] A. Hatami, K. Pahlavan, M. Heidari, and F. Akgul, "On rss and toa based indoor geolocation - a comparative performance evaluation," in *Wireless Communications and Networking Conference, 2006. WCNC 2006. IEEE*, vol. 4, 2006, pp. 2267–2272.

[68] S. Seidel and T. Rappaport, "914 mhz path loss prediction models for indoor wireless communications in multifloored buildings," *Antennas and Propagation, IEEE Transactions on*, vol. 40, no. 2, pp. 207–217, 1992.

[69] C. Feng, S. Valaee, and Z. Tan, "Multiple target localization using compressive sensing," in *Global Telecommunications Conference, 2009. GLOBECOM 2009. IEEE*, 2009, pp. 1–6.

[70] C. Feng, W. Au, S. Valaee, and Z. Tan, "Compressive sensing based positioning using rss of wlan access points," in *INFOCOM, 2010 Proceedings IEEE*, 2010, pp. 1–9.

[71] J. G. Proakis, *Digital Communications*, S. W. Director, Ed.   McGraw-Hill, 2001.

[72] X. chuan Liu, S. Zhang, Q. yuan Zhao, and X. kang Lin, "A real-time algorithm for fingerprint localization based on clustering and spatial diversity," in *Ultra Modern Telecommunications and Control Systems and Workshops (ICUMT), 2010 International Congress on*, 2010, pp. 74–81.

[73] E. Candes and M. Wakin, "An introduction to compressive sampling," *Signal Processing Magazine, IEEE*, vol. 25, no. 2, pp. 21–30, 2008.

[74] C. C. Nathan Blaunstein, *Radio Propagation and Adaptive Antennas for Wireless Communication Links: Terrestrial, Atmospheric and Ionospheric*, K. CHANG, Ed. Wiley, 2007.

[75] C. Wong, R. Klukas, and G. Messier, "Using wlan infrastructure for angle-of-arrival indoor user location," in *Vehicular Technology Conference, 2008. VTC 2008-Fall. IEEE 68th*, 2008, pp. 1–5.

[76] A. Hafiizh, S. Obote, and K. Kagoshima, "Doa-rssi multiple subcarrier indoor location estimation in mimo-ofdm wlan aps structure," *World Applied Sciences Journal* 7, pp. 182–18, 2009.

[77] R. Tingley and K. Pahlavan, "Space-time measurement of indoor radio propagation," *Instrumentation and Measurement, IEEE Transactions on*, vol. 50, no. 1, pp. 22–31, 2001.

[78] A. A. Ali, V. D. Nguyen, K. Kyamakya, and A. Omar, "First arrival detection based on channel estimation for positioning in wireless ofdm systems," in *14th European Signal Processing Conference (EUSIPCO 2006), Florence, Italy*, 2006.

[79] P. Uthansakul and M. Uthansakul, "A novel wlan positioning technique employing time delay of successful transmission," in *Next Generation Mobile Applications, Services and Technologies, 2008. NGMAST '08. The Second International Conference on*, 2008, pp. 105–110.

[80] ——, "Wlan positioning based on joint toa and rss characteristics," *International Journal of Electrical and Information Engineering*, vol. 3, no. 7, pp. 395–402, 2009.

[81] M. Heidari, N. Alsindi, and K. Pahlavan, "Udp identification and error mitigation in toa-based indoor localization systems using neural network architecture," *Wireless Communications, IEEE Transactions on*, vol. 8, no. 7, pp. 3597–3607, 2009.

[82] I. Gven, C.-C. Chong, F. Watanabe, and H. Inamura, "Nlos identification and weighted least-squares localization for uwb systems using multipath channel statistics," *EURASIP Journal on Advances in Signal Processing*, no. 36, Jan. 2008.

[83] G. H. Golub and C. F. V. Loan, *Matrix Computations*, 3rd ed.   Johns Hopkins University Press, Baltimore and London, 1996.

[84] A. H. Sayed and Yousef, "Wireless location," in *Encyclopedia of Telecommunications*, 2003.

[85] A. Mahmood and G. Gaderer, "Timestamping for ieee 1588 based clock synchronization in wireless lan," in *Precision Clock Synchronization for Measurement, Control and Communication, 2009. ISPCS 2009. International Symposium on*, 2009, pp. 1–6.

[86] T. Cooklev, J. Eidson, and A. Pakdaman, "An implementation of ieee 1588 over ieee 802.11b for synchronization of wireless local area network nodes," *Instrumentation and Measurement, IEEE Transactions on*, vol. 56, no. 5, pp. 1632–1639, 2007.

[87] R. Exel, G. Gaderer, and P. Loschmidt, "Localisation of wireless lan nodes using accurate tdoa measurements," in *Wireless Communications and Networking Conference (WCNC), 2010 IEEE*, 2010, pp. 1–6.

[88] T. Khanzada, "Wireless communication techniques for indoor positioning and tracking applications," Ph.D. dissertation, The University of Magdeburg, 2010.

[89] F. Gustafsson and F. Gunnarsson, "Positioning using time-difference of arrival measurements," in *Acoustics, Speech, and Signal Processing, 2003. Proceedings. (ICASSP '03). 2003 IEEE International Conference on*, vol. 6, 2003, pp. VI–553–6 vol.6.

[90] N. Yilmazer, K. Jinhwan, and T. Sarkar, "Utilization of a unitary transform for efficient computation in the matrix pencil method to find the direction of arrival," *Antennas and Propagation, IEEE Transactions on*, vol. 54, no. 1, pp. 175–181, Jan. 2006.

[91] N. Yilmazer, A. Seckin, and T. Sarkar, "Multiple snapshot direct data domain approach and esprit method for direction of arrival estimation," *Digital Signal Processing A Review Journal*, vol. 18, no. 4, pp. 561–567, July 2008.

[92] M. Khan and M. Tufail, "Comparative analysis of various matrix pencil methods for direction of arrival estimation," in *Image Analysis and Signal Processing (IASP), 2010 International Conference on*, 2010, pp. 496–501.

[93] ——, "Multiple snapshot beamspace matrix pencil method for direction of arrival estimation," in *Industrial Mechatronics and Automation (ICIMA), 2010 2nd International Conference on*, vol. 2, 2010, pp. 288–291.

[94] K.-C. Huarng and C.-C. Yeh, "A unitary transformation method for angle-of-arrival estimation," *Signal Processing, IEEE Transactions on*, vol. 39, no. 4, pp. 975–977, 1991.

[95] M. Speth, S. Fechtel, G. Fock, and H. Meyr, "Optimum receiver design for wireless broad-band systems using ofdm. i," *Communications, IEEE Transactions on*, vol. 47, no. 11, pp. 1668–1677, 1999.

[96] ——, "Optimum receiver design for ofdm-based broadband transmission .ii. a case study," *Communications, IEEE Transactions on*, vol. 49, no. 4, pp. 571–578, 2001.

[97] A. N. Gaber;, L. D. Khalaf;, and A. M. Mustafa, "Synchronization and cell search algorithms in 3gpp long term evolution systems (fdd mode)," *WSEAS TRANSACTIONS on COMMUNICATIONS*, vol. 11, no. 2, pp. 70–81, February 2012.

[98] T. Schmidl and D. Cox, "Robust frequency and timing synchronization for ofdm," *Communications, IEEE Transactions on*, vol. 45, no. 12, pp. 1613–1621, 1997.

[99] R. Schmidt, "Multiple emitter location and signal parameter estimation," *Antennas and Propagation, IEEE Transactions on*, vol. 34, no. 3, pp. 276–280, 1986.

[100] N. Dharamdial, R. Adve, and R. Farha, "Multipath delay estimations using matrix pencil," in *Wireless Communications and Networking, 2003. WCNC 2003. 2003 IEEE*, vol. 1, 2003, pp. 632–635 vol.1.

[101] R. Adve, T. K. Sarkar, O. Pereira-Filho, and S. Rao, "Extrapolation of time-domain responses from three-dimensional conducting objects utilizing the matrix pencil technique," *Antennas and Propagation, IEEE Transactions on*, vol. 45, no. 1, pp. 147–156, 1997.

[102] S. W. Lang and J. H. McClellan, "Frequency estimation with maximum entropy spectral estimators," *Acoustics, Speech and Signal Processing, IEEE Transactions on*, vol. 28, no. 6, pp. 716–724, 1980.

[103] D. Tufts and R. Kumaresan, "Estimation of frequencies of multiple sinusoids: Making linear prediction perform like maximum likelihood," *Proceedings of the IEEE*, vol. 70, no. 9, pp. 975–989, 1982.

[104] M.-A. Pallas and G. Jourdain, "Active high resolution time delay estimation for large bt signals," *Signal Processing, IEEE Transactions on*, vol. 39, no. 4, pp. 781–788, 1991.

[105] M. Wax and T. Kailath, "Detection of signals by information theoretic criteria," *Acoustics, Speech and Signal Processing, IEEE Transactions on*, vol. 33, no. 2, pp. 387–392, 1985.

[106] A. Ali, V. D. Nguyen, K. Kyamakya, and A. Omar, "Estimation of the channel-impulse-response length for adaptive ofdm systems based on information theoretic criteria," in *Vehicular Technology Conference, 2006. VTC 2006-Spring. IEEE 63rd*, vol. 4, 2006, pp. 1888–1892.

[107] S.-S. Jeng and C.-W. Tsung, "Multipath direction finding with frequency allocation subspace smoothing for an ofdm wireless communication system," in *Personal, Indoor and Mobile Radio Communications, 2005. PIMRC 2005. IEEE 16th International Symposium on*, vol. 4, 2005, pp. 2471–2475 Vol. 4.

[108] F. Penna and D. Cabric, "Cooperative doa-only localization of primary users in cognitive radio networks," *EURASIP Journal on Wireless Communications and Networking*, vol. 2013, no. 1, pp. 1–14, 2013. [Online]. Available: http://dx.doi.org/10.1186/1687-1499-2013-107

[109] M. F. Khan and M. Tufail, "3d modified unitary matrix pencil method with automatic grouping of unknown parameters of far field signals," *Digital Signal Processing*, vol. 23, p. 355363, 2013.

[110] J. Liang, "Joint azimuth and elevation direction finding using cumulant," *Sensors Journal, IEEE*, vol. 9, no. 4, pp. 390–398, 2009.

[111] S. Kikuchi, H. Tsuji, and A. Sano, "Pair-matching method for estimating 2-d angle of arrival with a cross-correlation matrix," *Antennas and Wireless Propagation Letters, IEEE*, vol. 5, no. 1, pp. 35–40, 2006.

[112] N. Tayem and H. Kwon, "L-shape 2-dimensional arrival angle estimation with propagator method," *Antennas and Propagation, IEEE Transactions on*, vol. 53, no. 5, pp. 1622–1630, 2005.

[113] J. Liang and D. Liu, "Joint elevation and azimuth direction finding using l-shaped array," *Antennas and Propagation, IEEE Transactions on*, vol. 58, no. 6, pp. 2136–2141, 2010.

[114] W. W. J. Liang, X. Zeng and H. Chen, "L-shaped array-based elevation and azimuth direction finding in the presence of mutual coupling," *Signal Processing*, vol. 91, p. 13191328, 2011.

[115] F.-J. Chen, C. Fung, C.-W. Kok, and S. Kwong, "Estimation of two-dimensional frequencies using modified matrix pencil method," *Signal Processing, IEEE Transactions on*, vol. 55, no. 2, pp. 718–724, 2007.

[116] H. L. V. Trees, *Optimum Array Processing: Part IV of Detection, Estimation, and Modulation Theory*. John Wiley & Sons, New York, 2002.

[117] A. Pages-Zamora, J. Vidal, and D. Brooks, "Closed-form solution for positioning based on angle of arrival measurements," in *Personal, Indoor and Mobile Radio Communications, 2002. The 13th IEEE International Symposium on*, vol. 4, 2002, pp. 1522–1526 vol.4.

[118] S. Nardone, A. G. Lindgren, and K. F.Gong, "Fundamental properties and performance of conventional bearings-only target motion analysis," *Automatic Control, IEEE Transactions on*, vol. 29, no. 9, pp. 775–787, Sep 1984.

[119] A. Broumandan, T. Lin, J. Nielsen, and G. Lachapelle, "Practical results of hybrid aoa/tdoa geo-location estimation in cdma wireless networks," in *Vehicular Technology Conference, 2008. VTC 2008-Fall. IEEE 68th*, 2008, pp. 1–5.

[120] Y. Fu and Z. Tian, "Cramer-rao bounds for hybrid toa/doa- based location estimation in sensor networks," *Letters, IEEE*, vol. 16, no. 8, pp. 655–658, 2009.

[121] A. Catovic and Z. Sahinoglu, "The cramer-rao bounds of hybrid toa/rss and tdoa/rss location estimation schemes," *Communications Letters, IEEE*, vol. 8, no. 10, pp. 626–628, Oct 2004.

[122] P. Stoica and N. Arye, "Music, maximum likelihood, and cramer-rao bound," *Acoustics, Speech and, IEEE Transactions on*, vol. 37, no. 5, pp. 720–741, 1989.

[123] J.-Y. Huang and Q. Wan, "Comments on "the cramer-rao bounds of hybrid toa/rss and tdoa/rss location estimation schemes"," *Communications Letters, IEEE*, vol. 11, no. 11, pp. 848–849, November 2007.

[124] Y. Shen and M. Win, "On the accuracy of localization systems using wideband antenn arrays," *Communications, IEEE Transactions on*, vol. 58, no. 1, pp. 270–280, 2010.

[125] S. Hara, D. Anzai, T. Yabu, K. Lee, T. Derham, and R. Zemek, "A perturbation analysis on the performance of toa and tdoa localization in mixed los/nlos environments," *Communications, IEEE Transactions on*, vol. 61, no. 2, pp. 679–689, 2013.

[126] *Agilent Technologies, N5182A MXG RF Vector Signal Generator, product brochure, 2009.*

[127] *L-com Global Connectivity, 2.4 GHz to 5.8 GHz 3 dBi TriBand Rubber Duck Antenna SMA-Male, product brochure.*

[128] *Agilent Technologies, N9010A EXA Vector Signal Analyzer, product brochure, 2009.*

[129] R. Hamila, J. Vesma, H. Vuolle, and M. Renfora, "Joint estimation of carrier phase and symbol timing using polynomial-based maximum likelihood technique," in *Universal Personal Communications, 1998. ICUPC '98. IEEE 1998 International Conference on*, vol. 1, Oct 1998, pp. 369–373 vol.1.

[130] S. A. F. Heinrich Meyr, Marc Moeneclaey, *Digital Communication Receivers: Synchronization, Channel Estimation, and Signal Processing.* John Wiley & Sons, 1998.

[131] *Ettus Research, USRP2, product brochure, 2011.*

[132] *National Instruments, NI USRP-2921, product brochure, 2012.*

[133] *Agilent Technologies, ENA E5071C Network Analyzer, product brochure, 2011.*

[134] *Agilent Technologies, 86100A Infiniium DCA the wideband Time Domain Transmission, product brochure, 2001.*

[135] J. Khodjaev, Y. Park, and A. S. Malik, "Survey of nlos identification and error mitigation problems in uwb-based positioning algorithms for dense environments," *Annals of Telecommunications*, vol. 65, no. 5-6, pp. 301–311, 2009.

[136] S. Gezici, H. Kobayashi, and H. Poor, "Nonparametric nonline-of-sight identification," in *Vehicular Technology Conference, 2003. VTC 2003-Fall. 2003 IEEE 58th*, vol. 4, 2003, pp. 2544–2548 Vol.4.

[137] Y.-H. Jo, J.-Y. Lee, D.-H. Ha, and S.-H. Kang, "Accuracy enhancement for uwb indoor positioning using ray tracing," in *Position, Location, And Navigation Symposium, IEEE/ION*, 2006, pp. 565–568.

[138] M. Haardt and J. Nossek, "Unitary esprit: how to obtain increased estimation accuracy with a reduced computational burden," *Signal Processing, IEEE Transactions on*, vol. 43, no. 5, pp. 1232–1242, 1995.

[139] S. Liu and G. Trenkler, "Hadamard, khatri-rao, kronecker and other matrix products," *International Journal of Information and Systems Sciences*, vol. 4, no. 1, pp. 160–177, 2008.

[140] D. R. Fokkema, G. L. G. Sleijpen, and H. A. V. der Vorst, "Jacobi-davidson style qr and qz algorithms for the reduction of matrix pencils," *SIAM J. SCI. COMPUT.*, vol. 20, no. 1, pp. 94–125, 1998.

yes I want morebooks!

Buy your books fast and straightforward online - at one of the world's fastest growing online book stores! Environmentally sound due to Print-on-Demand technologies.

Buy your books online at
www.get-morebooks.com

Kaufen Sie Ihre Bücher schnell und unkompliziert online – auf einer der am schnellsten wachsenden Buchhandelsplattformen weltweit! Dank Print-On-Demand umwelt- und ressourcenschonend produziert.

Bücher schneller online kaufen
www.morebooks.de

OmniScriptum Marketing DEU GmbH
Heinrich-Böcking-Str. 6-8
D - 66121 Saarbrücken
Telefax: +49 681 93 81 567-9

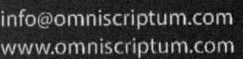
info@omniscriptum.com
www.omniscriptum.com

Printed by Books on Demand GmbH, Norderstedt / Germany